Making of America Project

Half Hours With the Telescope

Making of America Project

Half Hours With the Telescope

ISBN/EAN: 9783744675086

Printed in Europe, USA, Canada, Australia, Japan

Cover: Foto ©berggeist007 / pixelio.de

More available books at **www.hansebooks.com**

Map I

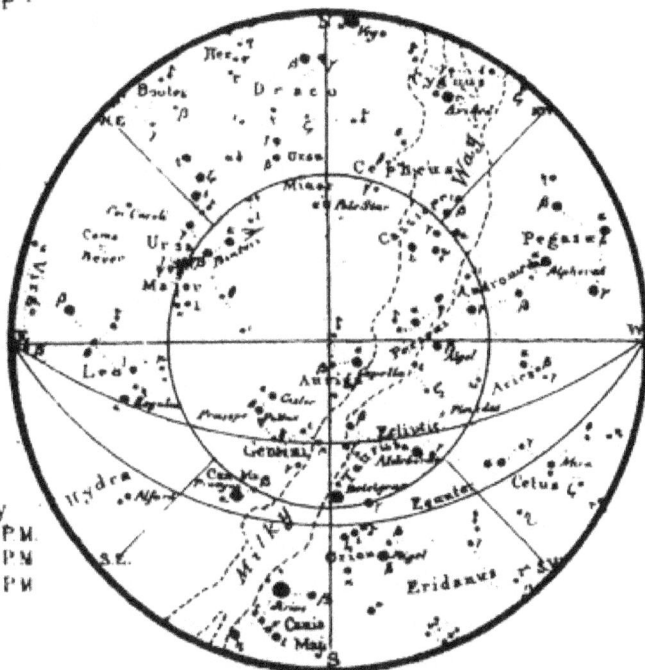

The Sky
Jan 20 10 PM
Feb 19 8 PM
Mar 21 6 PM

Hercules · Draco · Cygnus · Milky Way · Bootes · Ursa · Cepheus · Cor Caroli · Coma Berenices · Canes Venatici · Ursa Major · Cassiopeia · Polar Star · Pegasus · Andromeda · Alpheratz · Leo · Regulus · Auriga · Capella · Castor · Pollux · Algol · Arietis · Praesepe · Perseus · Pleiades · Cetus · Gemini · Milky Way · Equator · Eridanus · Orion · Rigel · Aries · Canis · Hydra · Alphard

N.E. · N.W. · S.E. · S.W. · S.

Map III.

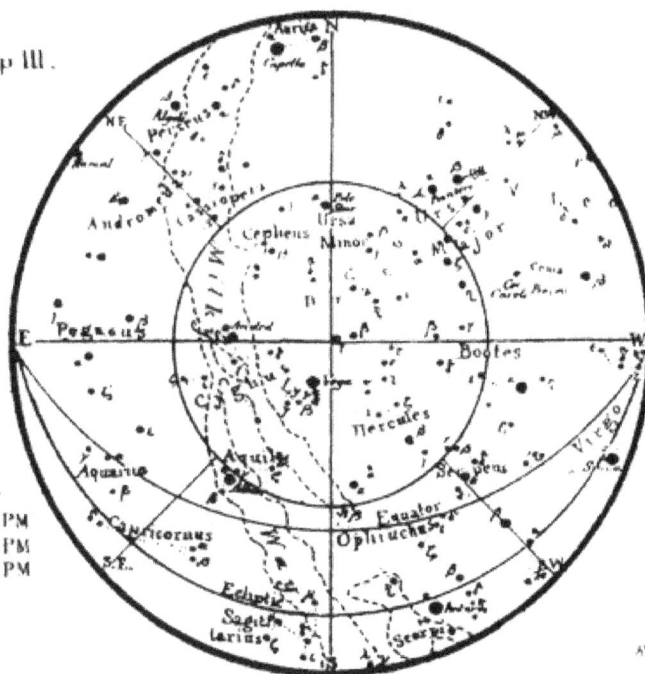

The Sky
Jul 22 10 PM
Aug 25 8 PM
Sep 23 6 PM

Auriga · Capella · Perseus · Algenib · Andromeda · Cassiopeia · Cepheus · Ursa Minor · Ursa Major · Leo · Cor Caroli Berenices · Pole Star · Milky Way · Draco · Bootes · Pegasus · Andromeda · Hercules · Corona Borealis · Equator · Ophiuchus · Aquila · Aquarius · Capricornus · Serpens · Virgo · Spica · Ecliptic · Sagittarius · Scorpio

E · N.E. · N.W. · W · S.E. · S.W.

PLATE 1.
Frontispiece
Map II.

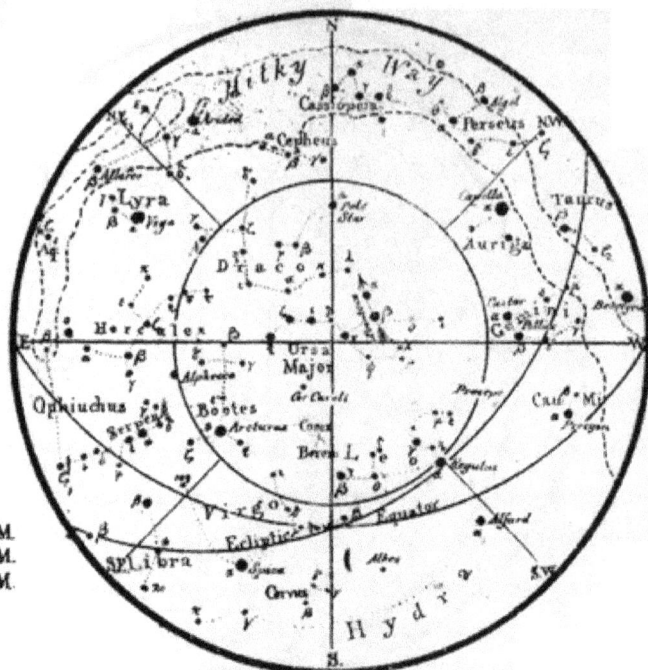

The Sky
Apr 20, 10 P.M.
May 21, 8 P.M.
Jun 22, 6 P.M.

Map IV.

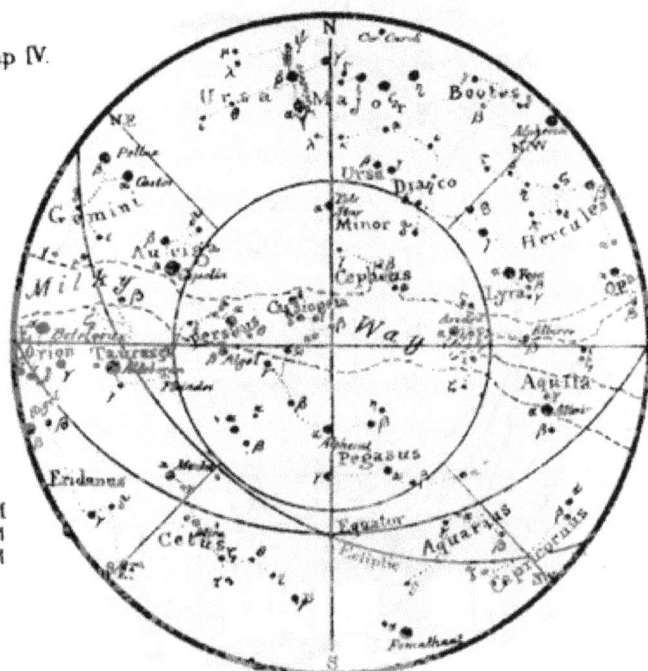

The Sky
Oct 23, 10 P.M.
Nov 22, 8 P.M.
Dec 21, 6 P.M.

HALF-HOURS

WITH

THE TELESCOPE;

BEING A POPULAR GUIDE TO THE USE OF THE TELESCOPE

AS A MEANS OF AMUSEMENT AND INSTRUCTION

BY

RICHARD A. PROCTOR, B.A., F.R.A.S.,

AUTHOR OF "SATURN AND ITS SYSTEM," ETC.

WITH ILLUSTRATIONS ON STONE AND WOOD.

An undevout astronomer is mad:
True, all things speak a God; but, in the small
Men trace out Him: in great He selzes man.
YOUNG.

NEW YORK:

G. P. PUTNAM'S SONS.

1873.

LONDON :

PRINTED BY WILLIAM CLOWES AND SONS, STAMFORD STREET
AND CHARING CROSS.

PREFACE.

—◆—

THE object which the Author and Publisher of this little work have proposed to themselves, has been the production, at a moderate price, of a useful and reliable guide to the amateur telescopist.

Among the celestial phenomena described or figured in this treatise, by far the larger number may be profitably examined with small telescopes, and there are none which are beyond the range of a good 3-inch achromatic.

The work also treats of the construction of telescopes, the nature and use of star-maps, and other subjects connected with the requirements of amateur observers.

R. A. P.

January, 1868.

CONTENTS.

CHAPTER I.

PAGE

A HALF-HOUR ON THE STRUCTURE OF THE TELESCOPE ...   1

CHAPTER II.

A HALF-HOUR WITH ORION, LEPUS, TAURUS, ETC. 33

CHAPTER III.

A HALF-HOUR WITH LYRA, HERCULES, CORVUS, CRATER, ETC. 47

CHAPTER IV.

A HALF-HOUR WITH BOOTES, SCORPIO, OPHIUCHUS, ETC. ... 56

CHAPTER V.

A HALF-HOUR WITH ANDROMEDA, CYGNUS, ETC. 66

CHAPTER VI.

HALF-HOURS WITH THE PLANETS 74

CHAPTER VII.

HALF-HOURS WITH THE SUN AND MOON 93

DESCRIPTION OF PLATES.

—◆◆◆—

PLATE I.—*Frontispiece.*

This plate presents the aspect of the heavens at the four seasons, dealt with in Chapters II., III., IV., and V. In each map of this plate the central point represents the point vertically over the observer's head, and the circumference represents his horizon. The plan of each map is such that the direction of a star or constellation, as respects the compass-points, and its elevation, also, above the horizon, at the given season, can be at once determined. Two illustrations of the use of the maps will serve to explain their nature better than any detailed description. Suppose first, that—at one of the hours named under Map I.—the observer wishes to find Castor and Pollux:—Turning to Map I. he sees that these stars lie in the lower left-hand quadrant, and very nearly towards the point marked S.E. ; that is, they are to be looked for on the sky towards the south-east. Also, it is seen that the two stars lie about one-fourth of the way from the centre towards the circumference. Hence, on the sky, the stars will be found about one-fourth of the way from the zenith towards the horizon : Castor will be seen immediately above Pollux. Next, suppose that at one of the hours named the observer wishes to learn what stars are visible towards the west and north-west :—Turning the map until the portion of the circumference marked W . . . . . . N.W. is lowermost, he sees that in the direction named the square of Pegasus lies not very high above the horizon, one diagonal of the square being vertical, the other nearly horizontal. Above the square is

Andromeda, to the right of which lies Cassiopeia, the stars
β and ε of this constellation lying directly towards the
north-west, while the star α lies almost exactly midway
between the zenith and the horizon. Above Andromeda,
a little towards the left, lies Perseus, Algol being almost
exactly towards the west and one-third of the way from the
zenith towards the horizon (because one-third of the way
from the centre towards the circumference of the map).
Almost exactly in the zenith is the star δ Aurigæ.

The four maps are miniatures of Maps I., IV., VII., and
X. of my 'Constellation Seasons,' fourth-magnitude stars,
however, being omitted.

PLATES II., III., IV., and V., illustrating Chapters II., III.,
IV., and V.

Plates II. and IV. contain four star-maps. They not only
serve to indicate the configuration of certain important star-
groups, but they illustrate the construction of maps, such as
the observer should make for himself when he wishes to
obtain an accurate knowledge of particular regions of the
sky. They are all made to one scale, and on the conical
projection—the simplest and best of all projections for maps
of this sort. The way in which the meridians and parallels
for this projection are laid down is described in my 'Handbook of the Stars.' With a little practice a few minutes will
suffice for sweeping out the equidistant circular arcs which
mark the parallels and ruling in the straight meridians.

The dotted line across three of the maps represents a
portion of the horizontal circle midway between the zenith
and the horizon at the hour at which the map is supposed
to be used. At other hours, of course, this line would be
differently situated.

Plates III. and V. represent fifty-two of the objects mentioned in the above-named chapters. As reference is made to
these figures in the text, little comment is here required. It
is to be remarked, however, that the circles, and especially

the small circles, do not represent the whole of the telescope's field of view, only a small portion of it. The object of these figures is to enable the observer to know what to expect when he turns his telescope towards a difficult double star. Many of the objects depicted are very easy doubles : these are given as objects of reference. The observer having seen the correspondence between an easy double and its picture, as respects the relation between the line joining the components and the apparent path of the double across the telescope's field of view, will know how to interpret the picture of a difficult double in this respect. And as all the small figures are drawn to one scale, he will also know how far apart he may expect to find the components of a difficult double. Thus he will have an exact conception of the sort of duplicity he is to look for, and this is—*crede experto*—a great step towards the detection of the star's duplicity.

PLATES VI. and VII., illustrating Chapters VI. and VII.

The views of Mercury, Venus, and Mars in these plates (except the smaller view of Jupiter in Plate VII.) are supposed to be seen with the same " power."

The observer must not expect to see the details presented in the views of Mars with anything like the distinctness I have here given to them. If he place the plate at a distance of six or seven yards he will see the views more nearly as Mars is likely to appear in a good three-inch aperture.

The chart of Mars is a reduction of one I have constructed from views by Mr. Dawes. I believe that nearly all the features included in the chart are permanent, though not always visible. I take this opportunity of noting that the eighteen orthographic pictures of Mars presented with my shilling chart are to be looked on rather as maps than as representing telescopic views. They illustrate usefully the varying presentation of Mars towards the earth. The observer can obtain other such illustrations for himself by filling in outlines, traced from those given at the foot of Plate VI.,

with details from the chart. It is to be noted that Mars
varies in presentation, not only as respects the greater or less
opening out of his equator towards the north or south, but as
respects the apparent slope of his polar axis to the right or
left. The four projections as shown, or inverted, or seen from
the back of the plate (held up to the light) give presentations
of Mars towards the sun at twelve periods of the Martial
year,—viz., at the autumnal and vernal equinoxes, at the
two solstices, and at intermediate periods corresponding to
our terrestrial months.

In fact, by means of these projections one might readily
form a series of sun-views of Mars resembling my 'Sun-views
of the Earth.'

In the first view of Jupiter it is to be remarked that the
three satellites outside the disc are supposed to be moving in
directions appreciably parallel to the belts on the disc—the
upper satellites from right to left, the lower one from left to
right. In general the satellites, when so near to the disc,
are not seen in a straight line, as the three shown in the
figure happen to be. Of the three spots on the disc, the
faintest is a satellite, the neighbouring dark spot its shadow,
the other dark spot the shadow of the satellite close to the
planet's disc.

HALF-HOURS WITH THE TELESCOPE.

CHAPTER I.

A HALF-HOUR ON THE STRUCTURE OF THE TELESCOPE.

THERE are few instruments which yield more pleasure and instruction than the Telescope. Even a small telescope—only an inch and a half or two inches, perhaps, in aperture—will serve to supply profitable amusement to those who know how to apply its powers. I have often seen with pleasure the surprise with which the performance even of an opera-glass, well steadied, and directed towards certain parts of the heavens, has been witnessed by those who have supposed that nothing but an expensive and colossal telescope could afford any views of interest. But a well-constructed achromatic of two or three inches in aperture will not merely supply amusement and instruction,—it may be made to do useful work.

The student of astronomy is often deterred from telescopic observation by the thought that in a field wherein so many have laboured, with abilities and means perhaps far surpassing those he may possess, he is little likely to reap results of any utility. He argues that, since the planets, stars, and nebulæ have been scanned by Herschel and Rosse, with their gigantic mirrors, and at Pulkova and Greenwich with refractors whose construction has taxed

B

to the utmost the ingenuity of the optician and
mechanic, it must be utterly useless for an unprac-
tised observer to direct a telescope of moderate
power to the examination of these objects.

Now, passing over the consideration that a small
telescope may afford its possessor much pleasure
of an intellectual and elevated character, even if
he is never able by its means to effect original
discoveries, two arguments may be urged in favour
of independent telescopic observation. In the first
place, the student who wishes to appreciate the
facts and theories of astronomy should familiarize
himself with the nature of that instrument to
which astronomers have been most largely indebted.
In the second place, some of the most important
discoveries in astronomy have been effected by
means of telescopes of moderate power used skil-
fully and systematically. One instance may suffice
to show what can be done in this way. The
well-known telescopist Goldschmidt (who commenced astronomical observation at the age of
forty-eight, in 1850) added fourteen asteroids to
the solar system, not to speak of important discoveries of nebulæ and variable stars, by means of
a telescope only five feet in focal length, mounted
on a movable tripod stand.

The feeling experienced by those who look
through a telescope for the first time,—especially
if it is directed upon a planet or nebula—is commonly one of disappointment. They have been
told that such and such powers will exhibit Jupiter's
belts, Saturn's rings, and the continent-outlines on
Mars; yet, though perhaps a higher power is
applied, they fail to detect these appearances, and
can hardly believe that they are perfectly distinct
to the practised eye.

The expectations of the beginner are especially

liable to disappointment in one particular. He forms an estimate of the view he is to obtain of a planet by multiplying the apparent diameter of the planet by the magnifying power of his telescope, and comparing the result with the apparent diameter of the sun or moon. Let us suppose, for instance, that on the day of observation Jupiter's apparent diameter is 45″, and that the telescopic power applied is 40, then in the telescope Jupiter should appear to have a diameter of 1800″, or half a degree, which is about the same as the moon's apparent diameter. But when the observer looks through the telescope he obtains a view—interesting, indeed, and instructive—but very different from what the above calculation would lead him to expect. He sees a disc apparently much smaller than the moon's, and not nearly so well-defined in outline; in a line with the disc's centre there appear three or four minute dots of light, the satellites of the planet; and, perhaps, if the weather is favourable and the observer watchful, he will be able to detect faint traces of belts across the planet's disc.

Yet in such a case the telescope is not in fault. The planet really appears of the estimated size. In fact, it is often possible to prove this in a very simple manner. If the observer wait until the planet and the moon are pretty near together, he will find that it is possible to view the planet with one eye through the telescope and the moon with the unaided eye, in such a manner that the two discs may coincide, and thus their relative apparent dimensions be at once recognised. Nor should the indistinctness and incompleteness of the view be attributed to imperfection of the telescope; they are partly due to the nature of the observation and the low power employed, and partly to the inexperience of the beginner.

It is to such a beginner that the following pages are specially addressed, with the hope of affording him aid and encouragement in the use of one of the most enchanting of scientific instruments,—an instrument that has created for astronomers a new sense, so to speak, by which, in the words of the ancient poet:

> Subjecere oculis distantia sidera nostris,
> Ætheraque ingenio supposuere suo.

In the first place, it is necessary that the beginner should rightly know what is the nature of the instrument he is to use. And this is the more necessary because, while it is perfectly easy to obtain such knowledge without any profound acquaintance with the science of optics, yet in many popular works on this subject the really important points are omitted, and even in scientific works such points are too often left to be gathered from a formula. When the observer has learnt what it is that his instrument is actually to do for him, he will know how to estimate its performance, and how to vary the application of its powers—whether illuminating or magnifying—according to the nature of the object to be observed.

Let us consider what it is that limits the range of *natural* vision applied to distant objects. What causes an object to become invisible as its distance increases? Two things are necessary that an object should be visible. It must be *large* enough to be appreciated by the eye, and it must *send light* enough. Thus increase of distance may render an object invisible, either through diminution of its apparent size, or through diminution in the quantity of light it sends to the eye, or through both these causes combined. A telescope, therefore, or (as its name implies) an instrument to render

distant objects visible, must be both a magnifying
and an illuminating instrument.

Let E F, fig. 1, be an object, not near to A B as
in the figure, but so far off that the bounding lines
from A and B would meet at the
point corresponding to the point P.
Then if a large convex glass A B
(called an *object-glass*) be interposed
between the object and the eye, all
those rays which, proceeding from
P, fall on A B, will be caused to
converge nearly to a point *p*. The
same is true for every point of the
object E M F, and thus a small
image, *e m f*, will be formed. This
image will not lie exactly on a flat
surface, but will be curved about
the point midway between A and
B as a centre. Now if the lens
A B is removed, and an eye is
placed at *m* to view the distant object E M F, those rays only from
each point of the object which
fall on the pupil of the eye (whose
diameter is about equal to *m p* suppose) will serve to render the object visible. On the other hand,
every point of the image *e m f* has
received the whole of the light
gathered up by the large glass A B.
If then we can only make this
light *available*, it is clear that we
shall have acquired a large in-
crease of *light* from the distant object. Now it will
be noticed that the light which has converged to
p, diverges from *p* so that an eye, placed that
this diverging pencil of rays may fall upon it,

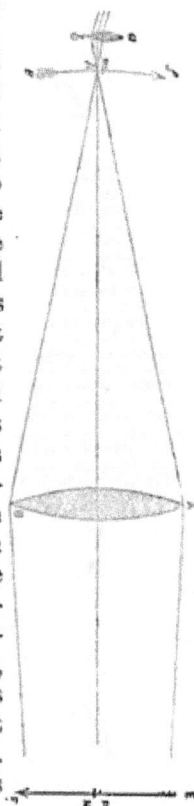

Fig. 1.

would be too small to receive the whole of
the pencil. Or, if it did receive the whole of this
pencil, it clearly could not receive the whole of
the pencils proceeding from other parts of the
image *e m f*. *Something* would be gained, though,
even in this case, since it is clear that an eye thus
placed at a distance of ten inches from *e m f* (which
is about the average distance of distinct vision)
would not only receive much more light from the
image *e m f*, than it would from the object E M F,
but see the image much larger than the object. It
is in this way that a simple object-glass forms a
telescope, a circumstance we shall presently have to
notice more at length. But we want to gain the
full benefit of the light which has been gathered up
for us by our object-glass. We therefore interpose
a small convex glass *a b* (called an eye-glass) between the image and the eye, at such a distance
from the image that the divergent pencil of rays
is converted into a pencil of parallel or nearly
parallel rays. Call this an emergent pencil. Then
all the emergent pencils now converge to a point
on the axial line *m* M (produced beyond *m*), and an
eye suitably placed can take in all of them at once.
Thus the whole, or a large part, of the image is seen
at once. But the image is seen inverted as shown.
This is the Telescope, as it was first discovered,
and such an arrangement would now be called a
simple astronomical Telescope.

Let us clearly understand what each part of the
astronomical telescope does for us :—

The object-glass A B gives us an illuminated
image, the amount of illumination depending on
the size of the object-glass. The eye-glass enables
us to examine the image microscopically.

We may apply eye-glasses of different focal
length. It is clear that the shorter the focal length

of *a b*, the nearer must *a b* be placed to the image, and the smaller will the emergent pencils be, but the greater the magnifying power of the eye-glass. If the emergent pencils are severally larger than the pupil of the eye, light is wasted at the expense of magnifying power. Therefore the eye-glass should never be of greater focal length than that which makes the emergent pencils about equal in diameter to the pupil of the eye. On the other hand, the eye-glass must not be of such small focal length that the image appears indistinct and contorted, or dull for want of light.

Let us compare with the arrangement exhibited in fig. 1 that adopted by Galileo. Surprise is sometimes expressed that this instrument, which in the hands of the great Florentine astronomer effected so much, should now be known as the *non-astronomical Telescope.* I think this will be readily understood when we compare the two arrangements.

In the Galilean Telescope a small concave eye-glass, *a b* (fig. 2), is placed between the object-glass and the image. In fact, no image is allowed to be formed in this arrangement, but the convergent pencils are intercepted by the concave eye-glass, and converted into parallel emergent pencils. Now in fig. 2 the concave eye-glass is so placed as to receive

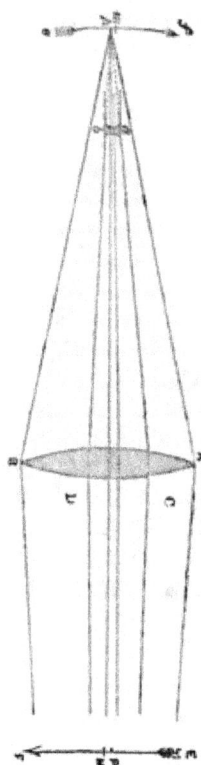

Fig. 2.

only a part of the convergent pencil A p B, and
this is the arrangement usually adopted.   By using
a concave glass of shorter focus, which would therefore be placed nearer to m p, the whole of the convergent pencil might be received in this as in the
former case.   But then the axis of the emergent
pencil, instead of returning (as we see it in fig. 1)
towards the axis of the telescope, would depart as
much *from* that axis.   Thus there would be no point
on the axis at which the eye could be so placed as
to receive emergent pencils showing any considerable
part of the object.   The difference may be compared
to that between looking through the small end of a
cone-shaped roll of paper and looking through the
large end; in the former case the eye sees at once all
that is to be seen through the roll (supposed fixed
in position), in the latter the eye may be moved
about so as to command the same range of view,
but *at any instant* sees over a much smaller range.

To return to the arrangement actually employed,
which is illustrated by the common opera-glass.   We
see that the full illuminating power of the telescope
is not brought into play.   But this is not the only
objection to the Galilean Telescope.   It is obvious
that if the part C D of the object-glass were covered,
the point P would not be visible, whereas, in the
astronomical arrangement no other effect is produced
on the visibility of an object, by covering part of the
object-glass, than a small loss of illumination.   In
other words, the dimensions of the field of view of
a Galilean Telescope depend on the size of the
object-glass, whereas in the astronomical Telescope
the field of view is independent of the size of the
object-glass.   The difference may be readily tested.
If we direct an opera-glass upon any object, we
shall find that any covering placed over a part of
the object-glass *becomes visible* when we look through

the instrument, interfering therefore *pro tanto* with the range of view. A covering similarly placed on any part of the object-glass of an astronomical telescope does not become visible when we look through the instrument. The distinction has a very important bearing on the theory of telescopic vision.

In considering the application of the telescope to practical observation, the circumstance that in the Galilean Telescope no real image is formed, is yet more important. A real image admits of measurement, linear or angular, while to a *virtual* image (such an image, for instance, as is formed by a common looking-glass) no such process can be applied. In simple observation the only noticeable effect of this difference is that, whereas in the astronomical Telescope a *stop* or diaphragm can be inserted in the tube so as to cut off what is called the *ragged edge* of the field of view (which includes all the part not reached by *full pencils of light* from the object-glass), there is no means of remedying the corresponding defect in the Galilean Telescope. It would be a very annoying defect in a telescope intended for astronomical observation, since in general the edge of the field of view is not perceptible at night. The unpleasant nature of the defect may be seen by looking through an opera-glass, and noticing the gradual fading away of light round the circumference of the field of view.

The properties of reflection as well as of refraction have been enlisted into the service of the astronomical observer. The formation of an image by means of a concave mirror is exhibited in fig. 3. As the observer's head would be placed between the object and the mirror, if the image, formed as in fig. 3, were to be microscopically examined, various devices are employed in the construction of reflecting telescopes to avoid the loss of light which would result—a loss which would be important even with the largest mirrors yet constructed. Thus, in Gregory's Telescope, a small mirror, having its concavity towards the great one, is placed in the axis of the tube and forms an image which is viewed through an aperture in the middle of the great mirror. A similar plan is adopted in Cassegrain's Telescope, a small convex mirror replacing the concave one. In Newton's Telescope a small inclined-plane reflector is used, which sends the pencil of light off at right-angles to the axis of the tube. In Herschel's Telescope the great mirror is inclined so that the image is formed at a slight distance from the axis of the telescope. In the two first cases the object is viewed in the usual or direct way, the image being erect in Gregory's and inverted in Cassegrain's. In the third the observer looks through the side of the telescope, seeing an inverted image of the object, In the last the observer sees the object inverted, but not altered as respects right and left. The last-mentioned method of viewing objects is the only one in which the observer's back is turned towards the object, yet this method is called the *front view*—apparently *quasi lucus a non lucendo.*

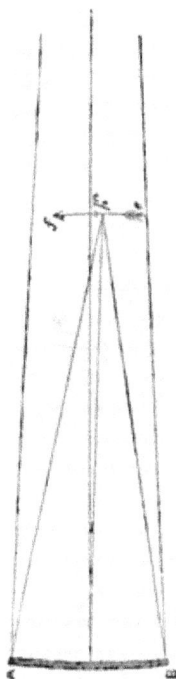

It appears, then, that in all astronomical Telescopes, reflecting or refracting, a *real image* of an object is submitted to microscopical examination.

Fig. 3.

Of this fact the possessor of a telescope may easily assure himself; for if the eye-glass be removed, and a small screen be placed at the focus of the object-glass, there will appear upon the screen a small picture of any object towards which the tube is turned. But the image may be viewed in another way which requires to be noticed. If the eye, placed at a distance of five or six inches from the image, be directed down the tube, the image will be seen as before; in fact, just as a single convex lens of short focus is the simplest microscope, so a simple convex lens of long focus is the simplest telescope.* But a singular circumstance will immediately attract the observer's notice. A real picture, or the image formed on the screen as in the former case, can be viewed at varying distances; but when we view the image directly, it will be found that for distinct vision the eye must be placed almost exactly at a fixed distance from the image. This peculiarity is more important than it might be thought at first sight. In fact, it is essential that the observer who would rightly apply the powers of his telescope, or fairly test its performance, should understand in what respect an image formed by an object-glass or object-mirror differs from a real object.

The peculiarities to be noted are the *curvature*, *indistinctness*, and *false colouring* of the image.

The curvature of the image is the least important of the three defects named—a fortunate circum-

* Such a telescope is most powerful with the shortest sight. It may be remarked that the use of a telescope often reveals a difference in the sight of the two eyes. In my own case, for instance, I have found that the left eye is very short-sighted, the sight of the right eye being of about the average range. Accordingly with my left eye a 5½-foot object-glass, alone, forms an effective telescope, with which I can see Jupiter's moons quite distinctly, and under favourable circumstances even Saturn's rings. I find that the full moon is too bright to be observed in this way without pain, except at low altitudes.

stance, since this defect admits neither of remedy nor modification. The image of a distant object, instead of lying in a plane, that is, forming what is technically called a *flat field*, forms part of a spherical surface whose centre is at the centre of the object-glass. Hence the centre of the field of view is somewhat nearer to the eye than are the outer parts of the field. The amount of curvature clearly depends on the extent of the field of view, and therefore is not great in powerful telescopes. Thus, if we suppose that the angular extent of the field is about half a degree (a large or low-power field), the centre is nearer than the boundary of the field by about 1-320th part only of the field's diameter.

The indistinctness of the image is partly due to the obliquity of the pencils which form parts of the image, and partly to what is termed *spherical aberration*. The first cause cannot be modified by the optician's skill, and is not important when the field of view is small. Spherical aberration causes those parts of a pencil which fall near the boundary of a convex lens to converge to a nearer (*i. e.* shorter) focus than those which fall near the centre. This may be corrected by a proper selection of the forms of the two lenses which replace, in all modern telescopes, the single lens hitherto considered.

The false colouring of the image is due to *chromatic aberration*. The pencil of light proceeding from a point, converges, not to one point, but to a short line of varying colour. Thus a series of coloured images is formed, at different distances from the object-glass. So that, if a screen were placed to receive the mean image *in focus*, a coloured fringe due to the other images (*out of focus, and therefore too large*) would surround the mean image.

Newton supposed that it was impossible to get rid of this defect, and therefore turned his attention to the construction of reflectors. But the discovery that the *dispersive* powers of different glasses are not proportional to their reflective powers, supplied opticians with the means of remedying the defect. Let us clearly understand what is the discovery referred to. If with a glass prism of a certain form we produce a spectrum of the sun, this spectrum will be thrown a certain distance away from the point on which the sun's rays would fall if not interfered with. This distance depends on the *refractive* power of the glass. The spectrum will have a certain length, depending on the *dispersive* power of the glass. Now, if we change our prism for another of exactly the same shape, but made of a different kind of glass, we shall find the spectrum thrown to a different spot. If it appeared that the length of the new spectrum was increased or diminished in exactly the same proportion as its distance from the line of the sun's direct light, it would have been hopeless to attempt to remedy chromatic aberration. Newton took it for granted that this was so. But the experiments of Hall and the Dollonds showed that there is no such strict proportionality between the dispersive and refractive powers of different kinds of glass. It accordingly becomes possible to correct the chromatic aberration of one glass by superadding that of another.

This is effected by combining, as shown in fig. 4, a convex lens of *crown* glass with a concave lens of *flint* glass, the convex lens being placed nearest to the object. A little colour still remains, but not enough to interfere seriously with the distinctness of the image.

Fig. 4.

But even if the image formed by the object-glass were perfect, yet this image, viewed

through a single convex lens of short focus placed as
in fig. 1, would appear curved, indistinct, coloured,
and also *distorted*, because
viewed by pencils of light
which do not pass through
the centre of the eye-glass.
These effects can be diminished (but not entirely removed *together*) by using an
eye-piece consisting of two lenses instead of a single
eye-glass. The two forms of eye-piece most commonly employed are exhibited in figs. 5 and 6. Fig.
5 is Huyghens' eye-piece, called also the *negative* eye-
piece, because a real image
is formed *behind* the *field-
glass* (the lens which lies
nearest to the object-glass).
Fig. 6 represents Rams-
den's eye-piece, called also
the *positive* eye-piece, because the real image formed
by the object-glass lies *in front of* the field-glass.

Fig. 5.

Fig. 6.

The course of a slightly oblique pencil through
either eye-piece is exhibited in the figures. The
lenses are usually plano-convex, the convexities
being turned towards the object-glass in the negative eye-piece, and towards each other in the positive
eye-piece. Coddington has shown, however, that
the best forms for the lenses of the negative eye-
piece are those shown in fig. 5.

The negative eye-piece, being achromatic, is commonly employed in all observations requiring distinct vision only. But as it is clearly unfit for
observations requiring micrometrical measurement,
or reference to fixed lines at the focus of the
object-glass, the positive eye-piece is used for these
purposes.

For observing objects at great elevations the diagonal eye-tube is often convenient. Its construction is shown in fig. 7. A B C is a totally reflecting prism of glass. The rays from the object-glass fall on the face A B, are totally reflected on the face B C, and emerge through the face A C. In using this eye-piece, it must be remembered that it lengthens the sliding eye-tube, which must therefore be thrust further in, or the object will not be seen in focus.

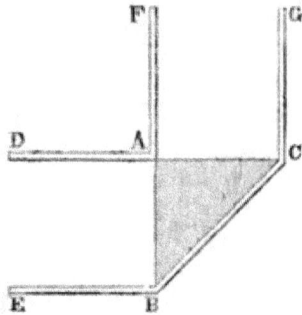

Fig. 7.

There is an arrangement by which the change of direction is made to take place between the two glasses of the eye-piece. With this arrangement (known as the *diagonal eye-piece*) no adjustment of the eye-tube is required. However, for amateurs' telescopes the more convenient arrangement is the diagonal eye-tube, since it enables the observer to apply any eye-piece he chooses, just as with the simple sliding eye-tube.

We come next to the important question of the *mounting* of our telescope.

The best known, and, in some respects, the simplest method of mounting a telescope for general observation is that known as the *altitude-and-azimuth* mounting. In this method the telescope is pointed towards an object by two motions,—one giving the tube the required *altitude* (or elevation), the other giving it the required *azimuth* (or direction as respects the compass points).

For small alt-azimuths the ordinary pillar-and-claw stand is sufficiently steady. For larger instru-

ments other arrangements are needed, both to give
the telescope steadiness, and to supply slow move-
ments in altitude and azimuth. The student will
find no difficulty in understanding the arrangement
of sliding-tubes and rack-work commonly adopted.
This arrangement seems to me to be in many
respects defective, however. The slow movement
in altitude is not uniform, but varies in effect according to the elevation of the object observed. It
is also limited in range; and quite a little series
of operations has to be gone through when it is
required to direct the telescope towards a new
quarter of the heavens. However expert the observer may become by practice in effecting these
operations, they necessarily take up some time (per-
formed as they must be in the dark, or by the light
of a small lantern), and during this time it often
happens that a favourable opportunity for observation is lost.

These disadvantages are obviated when the telescope is mounted in such a manner as is exhibited
in fig. 8, which represents a telescope of my own
construction. The slow movement in altitude is
given by rotating the rod $h\,e$, the endless screw in
which turns the small wheel at b, whose axle in
turn bears a pinion-wheel working in the teeth of
the quadrant a. The slow movement in azimuth is
given in like manner by rotating the rod $h'\,e'$, the
lantern-wheel at the end of which turns a crown-
wheel on whose axle is a pinion-wheel working in
the teeth of the circle c. The casings at e and e',
in which the rods $h\,e$ and $h'\,e'$ respectively work,
are so fastened by elastic cords that an upward
pressure on the handle h, or a downward pressure
on the handle h', at once releases the endless screw
or the crown-wheel respectively, so that the telescope can be swept at once through any desired

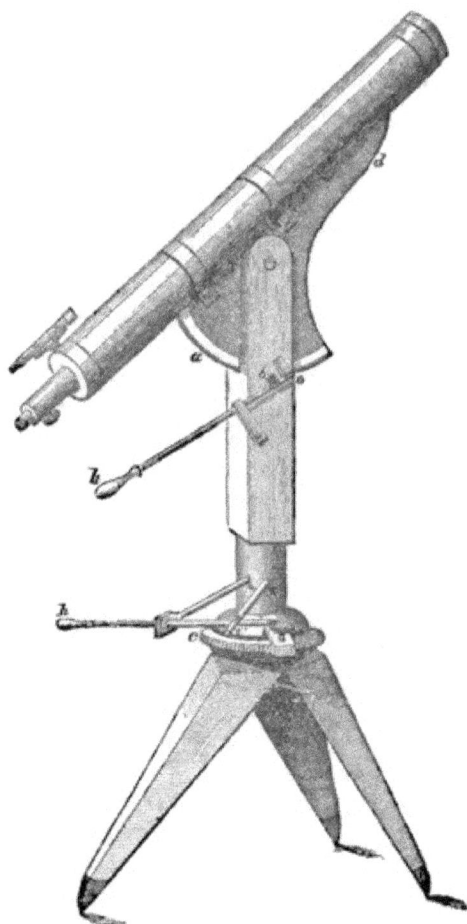

Fig. 8.

angle in altitude or azimuth. This method of
mounting has other advantages; the handles are
conveniently situated and constant in position; also,
as they do not work directly on the telescope,

c

they can be turned without setting the tube in vibration.

I do not recommend the mounting to be exactly as shown in fig. 8. That method is much too expensive for an alt-azimuth. But a simple arrangement of belted wheels in place of the toothed wheels a and c might very readily be prepared by the ingenious amateur telescopist; and I feel certain that the comfort and convenience of the arrangement would amply repay him for the labour it would cost him. My own telescope—though the large toothed-wheel and the quadrant were made inconveniently heavy (through a mistake of the workman who constructed the instrument)—worked as easily and almost as conveniently as an equatorial.

Still, it is well for the observer who wishes systematically to survey the heavens—and who can afford the expense—to obtain a well-mounted *equatorial*. In this method of mounting, the main axis is directed to the pole of the heavens; the other axis, at right angles to the first, carries the telescope-tube. One of the many methods adopted for mounting equatorials is that exhibited—with the omission of some minor details—in fig. 9. a is the polar axis, b is the axis (called the declination axis) which bears the telescope. The circles c and d serve to indicate, by means of verniers revolving with the axes, the motion of the telescope in right ascension and declination, respectively. The weight w serves to counterpoise the telescope, and the screws s, s, s, s, serve to adjust the instrument so that the polar axis shall be in its proper position. The advantage gained by the equatorial method of mounting is that only one motion is required to follow a star. Owing to the diurnal rotation of the earth, the stars appear to move uniformly in circles parallel to the celestial equator; and it is clear that

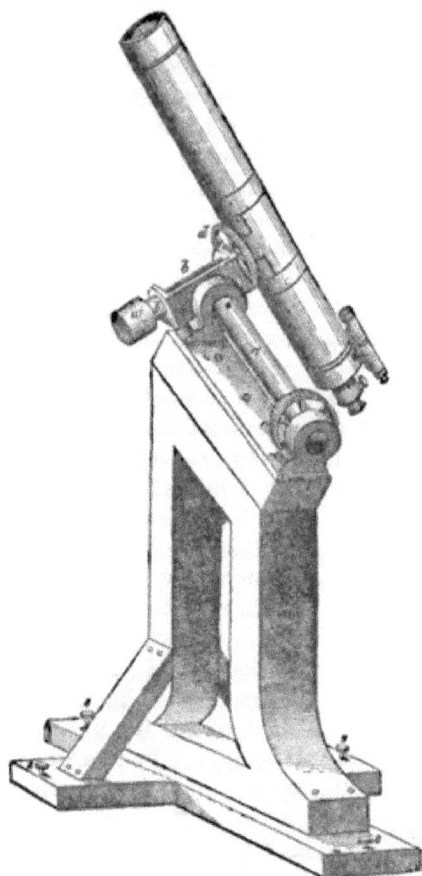

Fig. 9.

a star so moving will be kept in the field of view, if the telescope, once directed to the star, be made to revolve uniformly and at a proper rate round the polar axis.

The equatorial can be directed by means of the

c 2

circles c and d to any celestial object whose right
ascension and declination are known. On the other
hand, to bring an object into the field of view of an
alt-azimuth, it is necessary, either that the object
itself should be visible to the naked eye, or else
that the position of the object should be pretty
accurately learned from star-maps, so that it may be
picked up by the alt-azimuth after a little searching.
A small telescope called a *finder* is usually attached
to all powerful telescopes intended for general observation. The finder has a large field of view,
and is adjusted so as to have its axis parallel to
that of the large telescope. Thus a star brought
to the centre of the large field of the finder (indicated by the intersection of two lines placed at the
focus of the eye-glass) is at, or very near, the centre
of the small field of the large telescope.

If a telescope has no finder, it will be easy for
the student to construct one for himself, and will
be a useful exercise in optics. Two convex lenses
not very different in size from those shown in fig. 1,
and placed as there shown—the distance between
them being the sum of the focal lengths of the two
glasses—in a small tube of card, wood, or tin, will
serve the purpose of a finder for a small telescope.
It can be attached by wires to the telescope-tube,
and adjusted each night before commencing observation. The adjustment is thus managed:—a low
power being applied to the telescope, the tube is
turned towards a bright star; this is easily effected
with a low power; then the finder is to be fixed, by
means of its wires, in such a position that the star
shall be in the centre of the field of the finder when
also in the centre of the telescope's field. When
this has been done, the finder will greatly help the
observations of the evening; since with high powers
much time would be wasted in bringing an object

into the field of view of the telescope without the aid of a finder. Yet more time would be wasted in the case of an object not visible to the naked eye, but whose position with reference to several visible stars is known; since, while it is easy to bring the point required to the centre of the *finder's* field, in which the guiding stars are visible, it is very difficult to direct the *telescope's* tube on a point of this sort. A card tube with wire fastenings, such as we have described, may appear a very insignificant contrivance to the regular observer, with his well-mounted equatorial and carefully-adjusted finder. But to the first attempts of the amateur observer it affords no insignificant assistance, as I can aver from my own experience. Without it—a superior finder being wanting—our "half-hours" would soon be wasted away in that most wearisome and annoying of all employments, trying to "pick up" celestial objects.

It behoves me at this point to speak of star-maps. Such maps are of many different kinds. There are the Observatory maps, in which the places of thousands of stars are recorded with an amazing accuracy. Our beginner is not likely to make use of, or to want, such maps as these. Then there are maps merely intended to give a good general idea of the appearance of the heavens at different hours and seasons. Plate I. presents four maps of this sort; but a more complete series of eight-maps has been published by Messrs. Walton and Maberly in an octavo work; and my own 'Constellation-Seasons' give, at the same price, twelve quarto maps (of four of which those in Plate I. are miniatures), showing the appearance of the sky at any hour from month to month, or on any night, at successive intervals of two hours. But maps intermediate in character to these and to Observatory maps are required by the amateur observer. Such are the Society's six gnomonic maps, the set of six gnomonic maps in Johnstone's 'Atlas of Astronomy,' and my own set of twelve gnomonic maps. The Society's maps are a remarkably good set, containing on the scale of a ten-inch globe all the stars in the Catalogue of the Astronomical Society (down to the fifth magnitude). The distortion, however, is necessarily enormous when the celestial sphere is presented in only six gnomonic maps. In my maps all the stars of the British Association Catalogue down to the fifth magnitude are included on the scale of a six-inch globe. The distortion is scarcely a fourth of that in the Society's maps. The maps are so arranged that the relative positions of all the stars in each hemisphere can be readily gathered from a single view; and black duplicate-maps serve to show the appearance of the constellations.

It is often convenient to make small maps of a part of the heavens we may wish to study closely. My 'Handbook of the Stars' has been prepared to aid the student in the construction of such maps.

In selecting maps it is well to be able to recognise the amount of distortion and scale-variation. This may be done by examining the spaces included between successive parallels and meridians, near the edges and angles of the maps, and comparing these either with those in the centre of the map, or with the known figures and dimensions of the corresponding spaces on a globe.

We may now proceed to discuss the different tests which the intending purchaser of a telescope should apply to the instrument.

The excellence of an object-glass can be satisfactorily determined only by testing the performance of the telescope in the manner presently to be described. But it is well to examine the quality of

the glass as respects transparency and uniformity
of texture. Bubbles, scratches, and other such defects, are not very important, since they do not affect
the distinctness of the field as they would in a Galilean Telescope,—a little light is lost, and that is
all. The same remark applies to dust upon the
glass. The glass should be kept as free as possible
from dirt, damp, or dust, but it is not advisable to
remove every speck which, despite such precaution,
may accidentally fall upon the object-glass. When
it becomes necessary to clean the glass, it is to be
noted that the substance used should be soft, perfectly dry, and free from dust. Silk is often recom-
mended, but some silk is exceedingly objectionable
in texture,—old silk, perfectly soft to the touch, is
perhaps as good as anything. If the dust which
has fallen on the glass is at all gritty, the glass will
suffer by the method of cleaning commonly adopted,
in which the dust is *gathered up* by pressure. The
proper method is to clean a small space near the
edge of the glass, and to *sweep* from that space as
centre. In this way the dust is *pushed before* the
silk or wash-leather, and does not cut the glass. It
is well always to suspect the presence of gritty dust,
and adopt this cautious method of cleaning.

The two glasses should on no account be separated.

In examining an eye-piece, the quality of the
glass should be noted, and care taken that both
glasses (but especially the field-glass) are free from
the least speck, scratch, or blemish of any kind, for
these defects will be exhibited in a magnified state
in the field of view. Hence the eye-pieces require
to be as carefully preserved from damp and dust as
the object-glass, and to be more frequently cleaned.

The tube of the telescope should be light, but
strong, and free from vibration. Its quality in the
last respect can be tested by lightly striking it

when mounted; the sound given out should be dead or non-resonant. The inside of the tube must absorb extraneous light, and should therefore be coloured a dull black; and stops of varying radius should be placed along its length with the same object. Sliding tubes, rack-work, etc., should work closely, yet easily.

The telescope should be well balanced for vision with the small astronomical eye-pieces. But as there is often occasion to use appliances which disturb the balance, it is well to have the means of at once restoring equilibrium. A cord ring running round the tube (pretty tightly, so as to rest still when the tube is inclined), and bearing a small weight, will be all that is required for this purpose; it must be slipped along the tube until the tube is found to be perfectly balanced. Nothing is more annoying than, after getting a star well in the field, to see the tube shift its position through defective balance, and thus to have to search again for the star. Even with such an arrangement as is shown in fig. 8, though the tube cannot readily shift its position, it is better to have it well balanced.

The quality of the stand has a very important influence on the performance of a telescope. In fact, a moderately good telescope, mounted on a steady stand, working easily and conveniently, will not only enable the observer to pass his time much more pleasantly, but will absolutely exhibit more difficult objects than a finer instrument on a rickety, ill-arranged stand. A good observing-chair is also a matter of some importance, the least constraint or awkwardness of position detracting considerably from the power of distinct vision. Such, at least, is my own experience.

But the mere examination of the glasses, tube, mounting, &c., is only the first step in the series of tests which should be applied to a telescope, since the excellence of the instrument depends, not on its size, the beauty of its mounting, or any extraneous circumstances, but on its performance.

The observer should first determine whether the chromatic aberration is corrected. To ascertain this the telescope should be directed to the moon, or (better) to Jupiter, and accurately focussed for distinct vision. If, then, on moving the eye-piece towards the object-glass, a ring of purple appears round the margin of the object, and on moving the eye-glass in the contrary direction a ring of green, the chromatic aberration is corrected, since these are the colours of the secondary spectrum.

To determine whether the spherical aberration is corrected, the telescope should be directed towards a star of the third or fourth magnitude, and focussed for distinct vision. A cap with an aperture of about one-half its diameter should then be placed over the object-glass. If no new adjustment is required for distinct vision, the spherical aberration is corrected, since the mean focal length and the focal length of the central rays are equal. If, when the cap is on, the eye-piece has to be pulled out for distinct vision, the spherical aberration has not been fully corrected; if the eye-piece has to be pushed in, the aberration has been over-corrected. As a further test, we may cut off the central rays, by means of a circular card covering the middle of the object-glass, and compare the focal length for distinct vision with the focal length when the cap is applied. The extent of the spherical aberration may be thus determined; but if the first experiment gives a satisfactory result, no other is required.

A star of the first magnitude should next be brought into the field of view. If an irradiation from one side is perceived, part of the object-glass has not the same **refractive** power as the rest; and the part which is defective can be determined by applying in **different positions** a cap which hides half the object-glass. If the irradiation is double, it will probably **be** found that the object-glass has been too tightly screwed, and the defect will **disappear** when the glass is freed from such undue pressure.

If the object-glass is not quite at right angles to the axis of the tube, or if the eye-tube is at all inclined, a like irradiation will appear when a bright star is in the field. The former defect is not easily detected or remedied; nor is it commonly met with in the work of a careful optician. The latter defect may be detected by cutting out three circular cards of suitable size with a small aperture at the centre of each, and inserting one at each end of the eye-tube, and one over the object-glass. If the tube is rightly placed the apertures will of course lie in a right line, so that it will be possible to look through all three at once. If not, it will be easy to determine towards what part of the object-glass the eye-tube is directed, and to correct the position of the tube accordingly.

The best tests for determining the defining power of a telescope are close double or multiple stars, the components of which are not very unequal. The illuminating power should be tested by directing the telescope towards double or multiple stars having one or more minute components. Many of the nebulæ serve as tests both for illumination and defining power. As we proceed we shall meet with proper objects for testing different telescopes. For the present, let the following list suffice. It is selected from Admiral Smyth's tests, obtained by diminishing the aperture of a 6-in. telescope having a focal length of $8\frac{1}{2}$ feet:

A two-inch aperture, with powers of from 60 to 100, should exhibit

α Piscium (3″·5).	δ Cassiopeiæ (9″·5), mag. (4 and 7½).
γ Leonis (3″·2).	Polaris (18″·6), mag. (2½ and 9½).

A four-inch, powers 80 to 120, should exhibit

ξ Ursæ Majoris (2″·4).	σ Cassiopeiæ (3″·1), mag. (6 and 8).
γ Ceti (2″·6).	δ Geminorum (7″·1), mag. (4 and 9).

The tests in the first column are for definition, those in the second for illumination. It will be noticed that, though in the case of Polaris the smaller aperture may be expected to show the small star of less than the 9th magnitude, a larger aperture is required to show the 8th magnitude component of σ Cassiopeiæ, on account of the greater closeness of this double.

In favourable weather the following is a good general test of the performance of a telescope:— A star of the 3rd or 4th magnitude at a considerable elevation above the horizon should exhibit a small well defined disc, surrounded by two or three fine rings of light.

A telescope should not be mounted within doors, if it can be conveniently erected on solid ground, as every movement in the house will cause the instrument to vibrate unpleasantly. Further, if the telescope is placed in a warm room, currents of cold air from without will render observed objects hazy and indistinct. In fact, Sir W. Herschel considered that a telescope should not even be erected near a house or elevation of any kind round which currents of air are likely to be produced. If a telescope is used in a room, the temperature of the room should be made as nearly equal as possible to that of the outer air.

When a telescope is used out of doors a 'dew-cap,' that is, a tube of tin or pasteboard, some ten or twelve inches long, should be placed on the end of the instrument, so as to project beyond the object-glass. For glass is a good radiator of heat, so that dew falls heavily upon it, unless the radiation is in some way checked. The dew-cap does this effectually. It should be blackened within, especially if made of metal. "After use," says old Kitchener, "the telescope should be kept in a warm place long enough for any moisture on the object-glass to evaporate." If damp gets between the glasses it produces a fog (which opticians call a sweat) or even a seaweed-like vegetation, by which a valuable glass may be completely ruined.

The observer should not leave to the precious hours of the night the study of the bearing and position of the objects he proposes to examine. This should be done by day—an arrangement which has a twofold advantage,—the time available for observation is lengthened, and the eyes are spared sudden changes from darkness to light, and *vice versâ*. Besides, the eye is ill-fitted to examine difficult objects, after searching by candle-light amongst the minute details recorded in maps or globes. Of the effect of rest to the eye we have an instance in Sir J. Herschel's rediscovery of the satellites of Uranus, which he effected after keeping his eyes in darkness for a quarter of an hour. Kitchener, indeed, goes so far as to recommend (with a *crede experto*) an *interval of sleep* in the darkness of the observing-room before commencing operations. I have never tried the experiment, but I should expect it to have a bad rather than a good effect on the eyesight, as one commonly sees the eyes of a person who has been sleeping in his day-clothes look heavy and bloodshot.

The object or the part of an object to be observed should be brought as nearly as possible to the centre of the field of view. When there is no apparatus for keeping the telescope pointed upon an object, the best plan is so to direct the telescope by means of the finder, that the object shall be just out of the field of view, and be brought (by the earth's motion) across the centre of the field. Thus the vibrations which always follow the adjustment of the tube will have subsided before the object appears. The object should then be intently watched during the whole interval of its passage across the field of view.

It is important that the student should recognise the fact that the highest powers do not necessarily give the best views of celestial objects. High powers in all cases increase the difficulty of observation, since they diminish the field of view and the illumination of the object, increase the motion with which (owing to the earth's motion) the image moves across the field, and magnify all defects due to instability of the stand, imperfection of the object-glass, or undulation of the atmosphere. A good object-glass of three inches aperture will in very favourable weather bear a power of about 300, when applied to the observation of close-double or multiple stars, but for all other observations much lower powers should be used. Nothing but failure and annoyance can follow the attempt to employ the highest powers on unsuitable objects or in unfavourable weather.

The greatest care should be taken in focussing the telescope. When high powers are used this is a matter of some delicacy. It would be well if the eye-pieces intended for a telescope were so constructed that when the telescope is focussed for one, this might be replaced by any other without necessitating any use of the focussing rack-work. This could be readily effected by suitably placing the shoulder which limits the insertion of the eye-piece.

It will be found that, even in the worst weather for observation, there are instants of distinct vision (with moderate powers) during which the careful observer may catch sight of important details; and, similarly, in the best observing weather, there are moments of unusually distinct vision well worth patient waiting for, since in such weather alone the full powers of the telescope can be employed.

The telescopist should not be deterred from observation by the presence of fog or haze, since with a hazy sky definition is often singularly good.

The observer must not expect distinct vision of objects near the horizon. Objects near the eastern horizon during the time of morning twilight are especially confused by atmospheric undulations; in fact, early morning is a very unfavourable time for the observation of all objects.

The same rules which we have been applying to refractors, serve for reflectors. The performance of a reflector will be found to differ in some respects, however, from that of a refractor. Mr. Dawes is, we believe, now engaged in testing reflectors, and his unequalled experience of refractors will enable him to pronounce decisively on the relative merits of the two classes of telescopes.

We have little to say respecting the construction of telescopes. Whether it is advisable or not for an amateur observer to attempt the construction of his own telescope is a question depending entirely on his mechanical ability and ingenuity. My own experience of telescope construction is confined to the conversion of a 3-feet into a 5½-feet telescope. This operation involved some difficulties, since the aperture had to be increased by about an inch. I found a tubing made of alternate layers of card and calico well pasted together, to be both light and strong. But for the full length of tube I think a core of metal is wanted. A learned and ingenious friend, Mr. Sharp, Fellow of St. John's College, informs me that a tube of tin, covered with layers of brown paper, well pasted and thicker near the middle of the tube, forms a light and strong telescope-tube, almost wholly free from vibration.

Suffer no inexperienced person to deal with your object-glass. I knew a valuable glass ruined by the proceedings of a workman who had been told to attach three pieces of brass round the cell of the double lens. What he had done remained unknown, but ever after a wretched glare of light surrounded all objects of any brilliancy.

One word about the inversion of objects by the astronomical telescope. It is singular that any difficulty should be felt about so simple a matter, yet I have seen in the writings of more than one distinguished astronomer, wholly incorrect views as to the nature of the inversion. One tells us that to obtain the correct presentation from a picture taken with a telescope, the view should be inverted, held up to the light, and looked at from the back of the paper. Another tells us to invert the picture and hold it opposite a looking-glass. Neither method is correct. The simple correction wanted is to hold the picture upside down—the same change which brings the top to the bottom brings the right to the left, i. e., fully corrects the inversion.

In the case, however, of a picture taken by an

Herschelian reflector, the inversion not being complete, a different method must be adopted.   In fact, either of the above-named processes, incorrect for the ordinary astronomical, would be correct for the Herschelian Telescope.   The latter inverts but does not reverse right and left; therefore after inverting our picture we must interchange right and left because they have been reversed by the inversion. This is effected either by looking at the picture from behind, or by holding it up to a mirror.

PLATE II

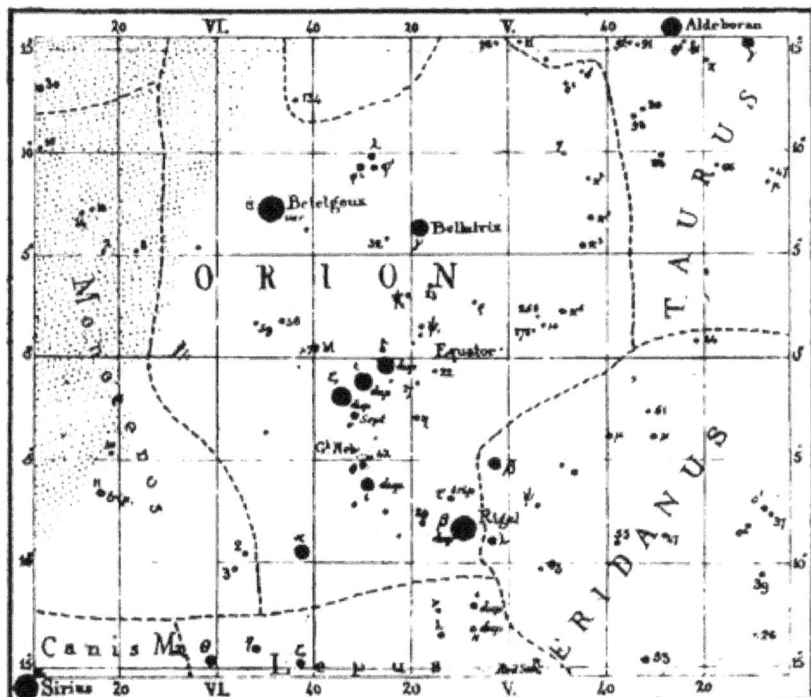

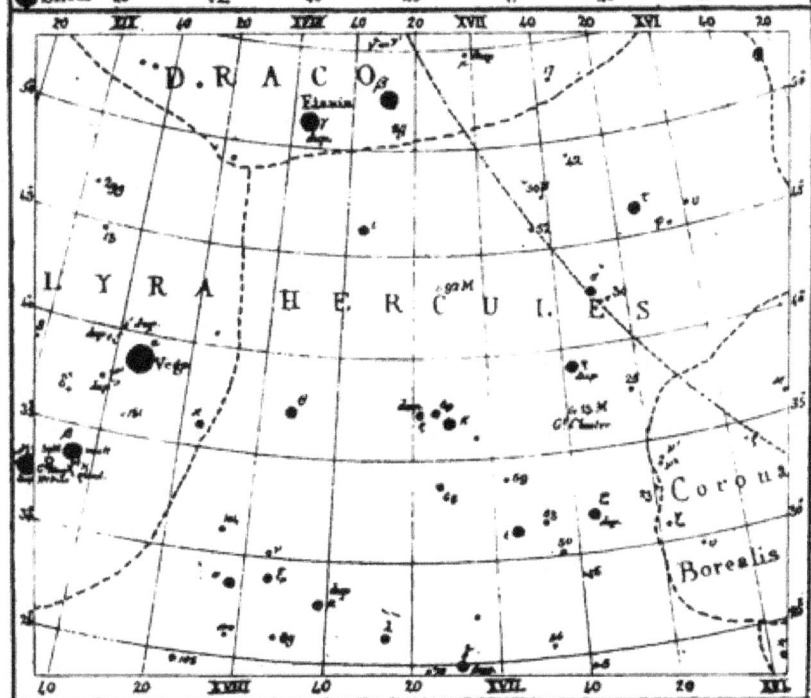

R. A. Proctor, del. et lith.

CHAPTER II.

A HALF-HOUR WITH ORION, LEPUS
TAURUS, ETC.

ANY of the half-hours here assigned to the constellation-seasons may be taken first, and the rest in
seasonal or cyclic order. The following introductory
remarks are applicable to each:—

If we stand on an open space, on any clear night,
we see above us the celestial dome spangled with
stars, apparently fixed in position. But after a little
time it becomes clear that these orbs are slowly
shifting their position. Those near the eastern
horizon are rising, those near the western setting.
Careful and continuous observation would show that
the stars are all moving in the same way, precisely,
as they would if they were fixed to the concave surface of a vast hollow sphere, and this sphere rotated
about an axis. This axis, in our latitude, is inclined
about $51\frac{1}{4}°$ to the horizon. Of course only one end
of this imaginary axis can be above our horizon.
This end lies very near a star which it will be well
for us to become acquainted with at the beginning
of our operations. It lies almost exactly towards
the north, and is raised from 50° to 53° (according
to the season and hour) above the horizon. There
is an easy method of finding it.

We must first find the Greater Bear. 'It will be
seen from Plate 1, that on a spring evening the seven
conspicuous stars of this constellation are to be
looked for towards the north-east, about half way
between the horizon and the point overhead (or

D

zenith), the length of the set of stars being vertical. On a summer's evening the Great Bear is nearly overhead. On an autumn evening he is towards the north-west, the length of the set of seven being somewhat inclined to the horizon. Finally, on a winter's evening, he is low down towards the north, the length of the set of seven stars being nearly in a horizontal direction.

Having found the seven stars, we make use of the pointers α and β (shown in Plate 1) to indicate the place of the Pole-star, whose distance from the pointer α is rather more than three times the distance of α from β.

Now stand facing the Pole-star. Then all the stars are travelling round that star *in a direction contrary to that in which the hands of a watch move*. Thus the stars below the pole are moving *towards the right*, those above the pole *towards the left*, those to the right of the pole *upwards*, those to the left of the pole *downwards*.

Next face the south. Then all the stars on our left, that is, towards the east, are rising slantingly towards the south; those due south are moving horizontally to the right, that is, towards the west; and those on our right are passing slantingly downwards towards the west.

It is important to familiarise ourselves with these motions, because it is through them that objects pass out of the field of view of the telescope, and by moving the tube in a proper direction we can easily pick up an object that has thus passed away, whereas if we are not familiar with the varying motions in different parts of the celestial sphere, we may fail in the attempt to immediately recover an object, and waste time in the search for it.

The consideration of the celestial motions shows how advantageous it is, when using an alt-azimuth,

to observe objects as nearly as possible **due south.**
Of course in many cases this is impracticable, because a phenomenon we wish to watch may occur
when an object is not situated near the **meridian.**
But in examining double stars there is in general no
reason for selecting objects inconveniently situated.
We can wait till they come round to the meridian,
and then observe them more comfortably. Besides,
most objects are higher, and therefore better seen,
when due south.

Northern objects, and especially those within the
circle of perpetual apparition, often culminate (that
is, cross the meridian, or north and south line) at
too great a height for comfortable vision.   In this
case we should observe them towards the east or
west, and remember that in the first case they are
rising, and in the latter they are setting, and that in
both cases they have also a motion from left to
right.

If we allow an object to pass right **across the field**
of view (the telescope being fixed), the apparent
direction of its motion is the exact reverse of **the**
true direction of the star's motion. This will **serve**
as a guide in shifting the alt-azimuth after a **star**
has passed out of the field of view.

The following technical terms must be explained.
That part of the field of view towards which the star
appears to move is called the *preceding* part of the
field, the opposite being termed the *following* part.
The motion for all stars, except those lying in an
oval space extending from the zenith to the pole of
the heavens, is more or less from right to left (in
the inverted field).   Now, if we suppose a star to
move along a diameter of the field so as to divide the
field into two semicircles, then in all cases in which
this motion takes places from right to left, that semicircle which contains the lowest point (apparently)

of the field is the *northern* half, the other is the *southern* half. Over the oval space just mentioned the reverse holds.

Thus the field is divided into four quadrants, and these are termed *north following* (*n. f.*), and *south following* (*s. f.*); *north preceding* (*n. p.*), and *south preceding* (*s. p.*). The student can have no difficulty in interpreting these terms, since he knows which is the following and which the preceding *semicircle*, which the northern and which the southern. In the figures of plates 3 and 5, the letters *n. f., n. p., &c.,* are affixed to the proper quadrants. It is to be remembered that the quadrants thus indicated are measured either way from the point and feather of the diametral arrows.

Next, of the apparent annual motion of the stars. This takes place in exactly the same manner as the daily motion. If we view the sky at eight o'clock on any day, and again at the same hour one month later, we shall find that at the latter observation (as compared with the former) the heavens appear to have rotated by the *twelfth part* of a complete circumference, and the appearance presented is precisely the same as we should have observed had we waited for two hours (the *twelfth part* of a day) on the day of the first observation.

Our survey of the heavens is supposed to be commenced during the first quarter of the year, at ten o'clock on the 20th of January, or at nine on the 5th of February, or at eight on the 19th of February, or at seven on the 6th of March, or at hours intermediate to these on intermediate days.

We look first for the Great Bear towards the north-east, as already described, and thence find the Pole-star; turning towards which we see, towards the right and downwards, the two guardians of the pole (β and γ Ursæ Minoris). Immediately under the Pole-star is the Dragon's Head, a conspicuous diamond of stars. Just on the horizon is Vega, scintillating brilliantly. Overhead is the brilliant Capella, near which the Milky Way is seen passing down to the horizon on either side towards the quarters S.S.E. and N.N.W.

For the present our business is with the southern heavens, however.

Facing the south, we see a brilliant array of stars, Sirius unmistakeably overshining the rest. Orion is shining in full glory, his leading brilliant, Betelgeuse* being almost exactly on the meridian, and also almost exactly half way between the horizon and the zenith. In Plate 2 is given a map of this constellation and its neighbourhood.

Let us first turn the tube on Sirius. It is easy to get him in the field without the aid of a finder. The search will serve to illustrate a method often useful when a telescope has no finder. Having taking out the eye-piece—a low-power one, suppose —direct the tube nearly towards Sirius. On looking through it, a glare of light will be seen within the tube. Now, if the tube be slightly moved about, the light will be seen to wax and wane, according as the tube is more or less accurately directed. Following these indications, it will be found easy to direct the tube, so that the object-glass shall appear *full of light.* When this is done, insert the eye-piece, and the star will be seen in the field.

But the telescope is out of focus, therefore we must turn the small focussing screw. Observe the

---

* Betelgeuse—commonly interpreted the Giant's Shoulder —*ibt-al-jauza.* The words, however, really signify, "the armpit of the central one," Orion being so named because he is divided centrally by the equator.

charming chromatic changes—green, and red, and
blue light, purer than the hues of the rainbow, scintillating and coruscating with wonderful brilliancy.
As we get the focus, the excursions of these light
flashes diminish until—if the weather is favourable
—the star is seen, still scintillating, and much
brighter than to the naked eye, but reduced to a
small disc of light, surrounded (in the case of so
bright a star as Sirius) with a slight glare. If after
obtaining the focus the focussing rackwork be still
turned, we see a coruscating image as before. In
the case of a very brilliant star these coruscations
are so charming that we may be excused for calling
the observer's attention to them. The subject is
not without interest and difficulty as an optical one.
But the astronomer's object is to get rid of all
these flames and sprays of coloured light, so that he
has very little sympathy with the admiration which
Wordsworth is said to have expressed for out-of-focus
views of the stars.

We pass to more legitimate observations, noticing
in passing that Sirius is a double star, the companion being of the tenth magnitude, and distant
about ten seconds from the primary. But our beginner is not likely to see the companion, which is
a very difficult object, owing to the overpowering
brilliancy of the primary.

Orion affords the observer a splendid field of research. It is a constellation rich in double and
multiple stars, clusters, and nebulæ. We will begin
with an easy object.

The star δ (Plate 3), or *Mintaka*, the uppermost
of the three stars forming the belt, is a wide double.
The primary is of the second magnitude, the secondary of the seventh, both being white.

The star α (*Betelgeuse*) is an interesting object, on
account of its colour and brilliance, and as one of

the most remarkable variables in the heavens. It was first observed to be variable by Sir John Herschel in 1836. At this period its variations were " most marked and striking." This continued until 1840, when the changes became " much less conspicuous. In January, 1849, they had recommenced, and on December 5th, 1852, Mr. Fletcher observed α Orionis brighter than Capella, and actually the largest star in the northern hemisphere." That a star so conspicuous, and presumably so large, should present such remarkable variations, is a circumstance which adds an additional interest to the results which have rewarded the spectrum-analysis of this star by Mr. Huggins and Professor Miller. It appears that there is decisive evidence of the presence in this luminary of many elements known to exist in our own sun; amongst others **are** found sodium, magnesium, calcium, iron, and bismuth. Hydrogen appears to be absent, or, more correctly, there are no lines in the star's spectrum corresponding to those of hydrogen in the solar spectrum. Secchi considers that there is evidence of an actual change in the spectrum of the star, an opinion in which Mr. Huggins does not coincide. In the telescope Betelgeuse appears as " a rich and brilliant gem." says Lassell, " a rich topaz, in hue and brilliancy differing from any that I have seen."

Turn next to β (**Rigel**), the brightest star below **the** belt. This is a very **noted** double, and will severely test our observer's telescope, if small. The components are well separated (see Plate 3), compared with many easier doubles; the secondary is also of the seventh magnitude, so that neither **as** respects closeness nor smallness of the secondary, is Rigel a difficult object. It is the combination of the two features which makes it a test-object. Kitchener says a $1\frac{3}{4}$-inch object-glass should show

Rigel double; in earlier editions of his work he gave $2\frac{1}{4}$-inches as the necessary aperture. Smyth mentions Rigel as a test for a 4-inch aperture, with powers of from 80 to 120. A 3-inch aperture, however, will certainly show the companion. Rigel is an orange star, the companion blue.

Turn next to λ the northernmost of the set of three stars in the head of Orion. This is a triple star, though an aperture of 3 inches will show it only as a double. The components are 5″ apart, the colours pale white and violet. With the full powers of a $3\frac{1}{2}$-inch glass a faint companion may be seen above λ.

The star ζ, the lowest in the belt, may be tried with a $3\frac{1}{2}$-inch glass. It is a close double, the components being nearly equal, and about $2\frac{1}{2}$″ apart (see Plate 3).

For a change we will now try our telescope on a nebula, selecting the great nebula in the Sword. The place of this object is indicated in Plate 2. There can be no difficulty in finding it since it is clearly visible to the naked eye on a moonless night—the only sort of night on which an observer would care to look at nebulæ. A low power should be employed.

The nebula is shown in Plate 3 as I have seen it with a 3-inch aperture. We see nothing of those complex streams of light which are portrayed in the drawings of Herschel, Bond, and Lassell, but enough to excite our interest and wonder. What is this marvellous light-cloud? One could almost imagine that there was a strange prophetic meaning in the words which have been translated "Canst thou loose the bands of Orion?" Telescope after telescope had been turned on this wonderful object with the hope of resolving its light into stars. But it proved intractable to Herschel's great reflector, to

Lassell's 2-feet reflector, to Lord Rosse's 3-feet reflector, and even partially to the great 6-feet reflector. Then we hear of its supposed resolution into stars, Lord Rosse himself writing to Professor Nichol, in 1846, " I may safely say there can be little, if any, doubt as to the resolvability of the nebula ;—all about the trapezium is a mass of stars, the rest of the nebula also abounding with stars, and exhibiting the characteristics of resolvability strongly marked."

It was decided, therefore, that assuredly the great nebula is a congeries of stars, and not a mass of nebulous matter as had been surmised by Sir W. Herschel. And therefore astronomers were not a little surprised when it was proved by Mr. Huggins' spectrum-analysis that the nebula consists of gaseous matter. How widely extended this gaseous universe may be we cannot say. The general opinion is that the nebulæ are removed far beyond the fixed stars. If this were so, the dimensions of the Orion nebula would be indeed enormous, far larger probably than those of the whole system whereof our sun is a member. I believe this view is founded on insufficient evidence, but this would not be the place to discuss the subject. I shall merely point out that the nebula occurs in a region rich in stars, and if it is not, like the great nebula in Argo, clustered around a remarkable star, it is found associated in a manner which I cannot look upon as accidental with a set of small-magnitude stars, and notably with the trapezium which surrounds that very remarkable black gap within the nebula. The fact that the nebula shares the proper motion of the trapezium appears inexplicable if the nebula is really far out in space beyond the trapezium. A very small proper motion of the trapezium (alone) would long since have destroyed the remarkable agreement in the position of the dark gap and the trapezium which has been noticed for so many years.

But whether belonging to our system or far beyond it, the great nebula must have enormous dimensions. A vast gaseous system it is, sustained by what arrangements or forces we cannot tell, nor can we know what purposes it subserves. Mr. Huggins' discovery that comets have gaseous nuclei, (so far as the two he has yet examined show) may suggest the speculation that in the Orion nebula we see a vast system of comets travelling in extensive orbits around nuclear stars, and so slowly as to exhibit for long intervals of time an unchanged figure. " But of such speculations" we may say with Sir J. Herschel "there is no end."

To return to our telescopic observations :—The trapezium affords a useful test for the light-gathering power of the telescope. Large instruments exhibit nine stars. But our observer may be well satisfied with his instrument and his eye-sight if he can see five with a 3½-inch aperture.* A good 3-inch glass shows four distinctly. But with smaller apertures only three are visible.

The whole neighbourhood of the great nebula will well repay research. The observer may sweep over it carefully on any dark night with profit. Above the nebula is the star-cluster 362 H. The star ι (double as shown in Plate 3) below the nebula is involved in a strong nebulosity. And in searching over this region we meet with delicate double, triple, and multiple stars, which make the survey interesting with almost any power that may be applied.

Above the nebula is the star σ, a multiple star.

* I have never been able to see more than four with a 3¾-inch aperture. I give a view of the trapezium as seen with an 8-inch equatorial.

To an observer with a good 3½-inch glass σ appears as an octuple star. It is well seen, however, as a fine multiple star with a smaller aperture. Some of the stars of this group appear to be variable.

The star ρ Orionis is an unequal, easy double, the components being separated by nearly seven seconds. The primary is orange, the smaller star smalt-blue (see Plate 3).

The middle star of the belt (ε) has a distant blue companion. This star, like ι, is nebulous. In fact, the whole region within the triangle formed by stars γ, κ, and β is full of nebulous double and multiple stars, whose aggregation in this region I do not consider wholly accidental.

We have not explored half the wealth of Orion, but leave much for future observation. We must turn, however, to other constellations.

Below Orion is Lepus, the Hare, a small constellation containing some remarkable doubles. Among these we may note ξ, a white star with a scarlet companion; γ, a yellow and garnet double; and ι, a double star, white and pale violet, with a distant red companion. The star κ Leporis is a rather close double, white with a small green companion. The intensely red star R Leporis (a variable) will be found in the position indicated in the map.

Still keeping within the boundary of our map, we may next turn to the fine cluster 2 H (vii.) in Monoceros. This cluster is visible to the naked eye, and will be easily found. The nebula 2 H (iv.) is a remarkable one with a powerful telescope.

The star 11 Monocerotis is a fine triple star described by the elder Herschel as one of the finest sights in the heavens. Our observer, however, will see it as a double (see Plate 3). δ Monocerotis is an easy double, yellow and lavender.

We may now leave the region covered by the

map and take a survey of the heavens for some
objects well seen at this season.

Towards the south-east, high up above the
horizon, we see the twin-stars Castor and Pollux.
The upper is Castor, the finest double star visible
in the northern heavens. The components are
nearly equal and rather more than 5″ apart (see
Plate 3). Both are white according to the best
observers, but the smaller is thought by some to be
slightly greenish.

Pollux is a coarse but fine triple star (in large
instruments multiple). The components orange,
grey, and lilac.

There are many other fine objects in Gemini, but
we pass to Cancer.

The fine cluster Præsepe in Cancer may easily be
found as it is distinctly visible to the naked eye in
the position shown in Plate 1, Map I. In the
telescope it is seen as shown in Plate 3.

The star ι Cancri is a wide double, the colours
orange and blue.

Procyon, the first-magnitude star between Præsepe and Sirius, is finely coloured—yellow with a
distant orange companion, which appears to be
variable.

Below the Twins, almost in a line with them, is the
star α Hydræ, called Al Fard, or " the Solitary One."
It is a 2nd magnitude variable. I mention it, however, not on its own account, but as a guide to the
fine double ε Hydræ. This star is the middle one
of a group of three, lying between Pollux and Al
Fard rather nearer the latter. The components of
ε Hydræ are separated by about $3\frac{1}{2}″$ (see Plate 3).
The primary is of the fourth, the companion of
the eighth magnitude; the former is yellow, the
latter a ruddy purple. The period of ε Hydræ is
about 450 years.

The constellation Leo Minor, now due east and about midway between the horizon and the zenith, is well worth sweeping over. It contains several fine fields.

Let us next turn to the western heavens. Here there are some noteworthy objects.

To begin with, there are the Pleiades, showing to the naked eye only six or seven stars. In the telescope the Pleiades appear as shown in Plate 3.

The Hyades also show some fine fields with low powers.

Aldebaran, the principal star of the Hyades, as also of the constellation Taurus, is a noted red star. It is chiefly remarkable for the close spectroscopic analysis to which it has been subjected by Messrs. Huggins and Miller. Unlike Betelgeuse, the spectrum of Aldebaran exhibits the lines corresponding to hydrogen, and no less than eight metals—sodium, magnesium, calcium, iron, bismuth, tellurium, antimony, and mercury, are proved to exist in the constitution of this brilliant red star.

On the right of Aldebaran, in the position indicated in Plate 1, Map I., are the stars ζ and β Tauri. If with a low power the observer sweep from ζ towards β, he will soon find—not far from ζ (at a distance of about one-sixth of the distance separating β from ζ), the celebrated Crab nebula, known as 1 M. This was the first nebula discovered by Messier, and its discovery led to the formation of his catalogue of 103 nebulæ. In a small telescope this object appears as a nebulous light of oval form, no traces being seen of the wisps and sprays of light presented in Lord Rosse's well known picture of the nebula.

Here I shall conclude the labours of our first half-hour among the stars, noticing that the examination of Plate 1 will show what other constella-

tions besides those here considered are well situated
for observation at this season.   It will be remarked
that many constellations well seen in the third half-
hour (Chapter IV.) are favourably seen in the first
also, and *vice versâ*.   For instance, the constellation
Ursa Major well-placed towards the north-east in
the first quarter of the year, is equally well-placed
towards the north-west in the third, and similarly
of the constellation Cassiopeia.   The same relation
connects the second and fourth quarters of the
year.

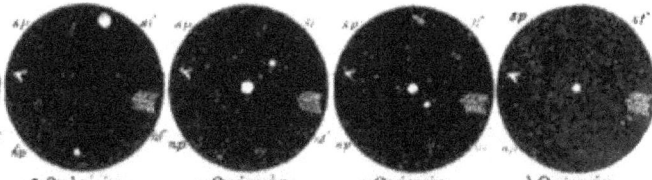

8 Orionis   ι Orionis   ς Orionis   λ Orionis

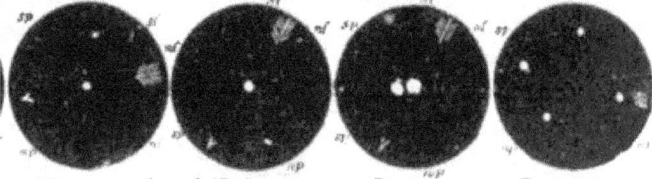

Monocerotis   ξ Hydræ   Castor   Trapezium

Orion   Pleiades   Præsepe

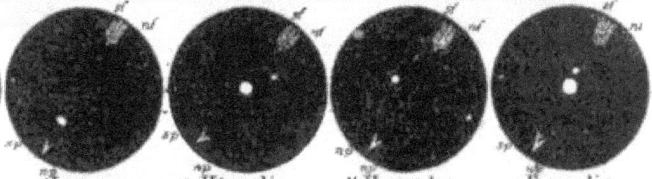

ξ Lyræ   δ Herculis   γ Herculis   α Herculis

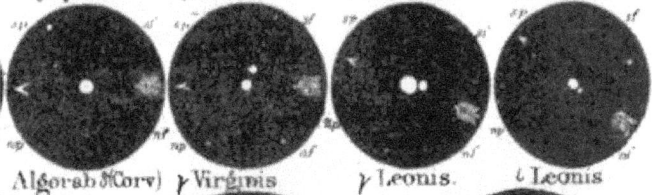

Algorab δ(Corv)   γ Virginis   γ Leonis   ι Leonis

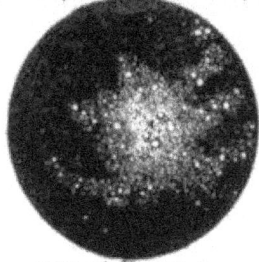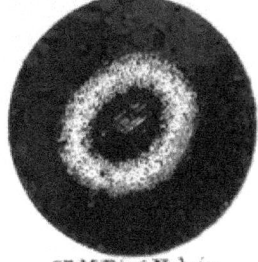

Lyræ   13 M Gt Cluster in Hercules   57 M Ring Neb in Lyra

CHAPTER III.

A HALF-HOUR WITH LYRA, HERCULES, CORVUS, CRATER, ETC.

THE observations now to be commenced are supposed to take place during the second quarter of the year,—at ten o'clock on the 20th of April, or at nine on the 5th of May, or at eight on the 21st of May, or at seven on the 5th of June, or at hours intermediate to these on intermediate days.

We again look first for the Great Bear, now near the zenith, and thence find the Pole-star. Turning towards the north, we see Cassiopeia between the Pole-star and the horizon. Towards the north-west is the brilliant Capella, and towards the north-east the equally brilliant Vega, beneath which, and somewhat northerly, is the cross in Cygnus. The Milky Way passes from the eastern horizon towards the north (low down), and so round to the western horizon.

In selecting a region for special observation, we shall adopt a different plan from that used in the preceding "half-hour." The region on the equator and towards the south is indeed particularly interesting, since it includes the nebular region in Virgo. Within this space nebulæ are clustered more closely than over any corresponding space in the heavens, save only the greater Magellanic cloud. But to the observer with telescopes of moderate power these nebulæ present few features of special interest; and there are regions of the sky now well situated for observation, which, at most other epochs are either

low down towards the horizon or inconveniently
near to the zenith. We shall therefore select one
of these, the region included in the second map of
Plate 2, and the neighbouring part of the celestial
sphere.

At any of the hours above named, the constellation Hercules lies towards the east. A quadrant
taken from the zenith to the eastern horizon passes
close to the last star (η) of the Great Bear's tail,
through β, a star in Boötes' head, near β Herculis,
between the two "Alphas" which mark the heads
of Hercules and Ophiuchus, and so past β Ophiuchi,
a third-magnitude star near the horizon. And here
we may turn aside for a moment to notice the
remarkable vertical row of six conspicuous stars
towards the east-south-east; these are, counting
them in order from the horizon, ζ, ϵ, and δ Ophiuchi,
ϵ, a, and δ Serpentis.

Let the telescope first be directed towards Vega.
This orb presents a brilliant appearance in the telescope. Its colour is a bluish-white. In an ordinary telescope Vega appears as a single star, but
with a large object-glass two distant small companions are seen. A nine-inch glass shows also
two small companions within a few seconds of Vega.
In the great Harvard refractor Vega is seen with
no less than thirty-five companions. I imagine
that all these stars, and others which can be
seen in neighbouring fields, indicate the association of Vega with the neighbouring stream of the
Milky Way.

Let our observer now direct his telescope to the
star ϵ Lyræ. Or rather, let him first closely examine this star with the naked eye. The star is
easily identified, since it lies to the left of Vega,
forming with ζ a small equilateral triangle. A careful scrutiny suffices to indicate a peculiarity in this

star. If our observer possesses very good eye-sight,
he will distinctly recognise it as a "naked-eye
double"; but more probably he will only notice
that it appears lengthened in a north and south
direction.*   In the finder the star is easily divided.
Applying a low power to the telescope itself, we
see ε Lyræ as a wide double, the line joining the
components lying nearly north and south.   The
southernmost component (the upper in the figure)
is called ε¹, the other ε².   Seen as a double, both
components appear white.

Now, if the observer's telescope is sufficiently
powerful, each of the components may be seen to
be itself double.   First try ε¹, the northern pair.
The line joining the components is directed as
shown in Plate 3.   The distance between them is
3"·2, their magnitudes 5 and 6½, and their colours
yellow and ruddy.   If the observer succeeds in
seeing ε¹ fairly divided, he will probably not fail
in detecting the duplicity of ε², though this is a
rather closer pair, the distance between the components being only 2"·6.   The magnitudes are 5
and 5½, both being white.   Between ε¹ and ε² are
three faint stars, possibly forming with the quadruple a single system.

Let us next turn to the third star of the equilateral triangle mentioned above—viz. to the star
ζ Lyræ.   This is a splendid but easy double.   It
is figured in Plate 3, but it must be noticed that

---

* Sir W. Herschel several times saw ε Lyræ as a double.
Bessel also relates that when he was a lad of thirteen he
could see this star double.   I think persons having average
eye-sight could see it double if they selected a suitable
hour for observation.   My own eye-sight is not good enough
for this, but I can distinctly see this star wedged whenever
the line joining the components is inclined about 45° to the
horizon, and also when Lyra is near the zenith.

E

the figure of ζ and the other nine small figures are
not drawn on the same scale as ϵ Lyræ. The actual
distance between the components of ζ Lyra is
44", or about one-fourth of the distance separating
ϵ^1 from ϵ^2. The components of ζ are very nearly
equal in magnitude, in colour topaz and green, the
topaz component being estimated as of the fifth magnitude, the green component intermediate between
the fifth and sixth magnitudes.

We may now turn to a star not figured in the
map, but readily found. It will be noticed that
the stars ϵ, a, β, and γ form, with two small stars
towards the left, a somewhat regular hexagonal
figure—a hexagon, in fact, having three equal long
sides and three nearly equal short sides alternating
with the others. The star η Lyræ forms the angle
next to ϵ. It is a wide and unequal double, as
figured in Plate 3. The larger component is azure
blue; the smaller is violet, and, being only of the
ninth magnitude, is somewhat difficult to catch with
apertures under 3 inches.

The star δ^2 Lyræ is orange, δ^1 blue. In the same
field with these are seen many other stars.

The stars γ^1 and γ^2 may also be seen in a single
field, the distance between them being about half
the moon's mean diameter. The former is quadruple,
the components being yellow, bluish, pale blue, and
blue.

Turn next to the stars β and γ Lyræ, the former
a multiple, the latter an unequal double star. It is
not, however, in these respects that these stars are
chiefly interesting, but for their variability. The
variability of γ has not indeed been fully esta-
blished, though it is certain that, having once been
less bright, γ is now considerably brighter than β.
The change, however, may be due to the variation of
β alone. This star is one of the most remarkable

variables known. Its period is 12d. 21h. 53m. 10s.
In this time it passes from a maximum brilliancy—
that of a star of the 3·4 magnitude—to a minimum
lustre equal to that of a star of the 4·3 magnitude,
thence to the same maximum brilliancy as before,
thence to another minimum of lustre—that of a star
of the 4·5 magnitude—and so to its maximum lustre
again, when the cycle of changes recommences.
These remarkable changes seem to point to the existence of two unequal dark satellites, whose dimensions bear a much greater proportion to those of
the bright components of β Lyræ than the dimensions of the members of the Solar System bear to
those of the sun. In this case, at any rate, the
conjecture hazarded about Algol, that the star re-
volves around a dark central orb, would be insufficient to account for the observed variation.

Nearly midway between β and γ lies the wonderful ring-nebula 57 M, of which an imperfect idea
will be conveyed by the last figure of Plate 3. This
nebula was discovered in 1772, by Darquier, at
Toulouse. It is seen as a ring of light with very
moderate telescopic power. In a good 3½-inch telescope the nebula exhibits a mottled appearance and
a sparkling light. Larger instruments exhibit a
faint light within the ring; and in Lord Rosse's
great Telescope "wisps of stars" are seen within,
and faint streaks of light stream from the outer
border of the ring. This nebula has been subjected
to spectrum-analysis by Mr. Huggins. It turns out
to be a gaseous nebula! In fact, ring-nebulæ—of
which only seven have been detected—seem to belong to the same class as the planetary nebulæ, all
of which exhibit the line-spectrum indicative of
gaseity. The brightest of the three lines seen in
the spectrum of the ring-nebula in Lyra presents
a rather peculiar appearance, "since it consists,"

says Mr. Huggins, "of two bright dots, corresponding to sections of the ring, and between these there is not darkness, but an excessively faint line joining them. This observation makes it probable that the faint nebulous matter occupying the central portion is similar in constitution to that of the ring."

The constellation Hercules also contains many very interesting objects. Let us first inspect a nebula presenting a remarkable contrast with that just described. I refer to the nebula 13 M, known as Halley's nebula (Plate 3). This nebula is visible to the naked eye, and in a good telescope it is a most wonderful object : "perhaps no one ever saw it for the first time without uttering a shout of wonder." It requires a very powerful telescope completely to resolve this fine nebula, but the outlying streamers may be resolved with a good 3-inch telescope. Sir W. Herschel considered that the number of the stars composing this wonderful object was at least 14,000. The accepted views respecting nebulæ would place this and other clusters far beyond the limits of our sidereal system, and would make the component stars not very unequal (on the average) to our own sun. It seems to me far more probable, on the contrary, that the cluster belongs to our own system, and that its components are very much smaller than the average of separate stars. Perhaps the whole mass of the cluster does not exceed that of an average first-magnitude star.

The nebulæ 92 M and 50 H may be found, after a little searching, from the positions indicated in the map. They are both well worthy of study, the former being a very bright globular cluster, the latter a bright and large round nebula. The spectra of these, as of the great cluster, resemble the solar spectrum, being continuous, though, of course, very much fainter.

The star δ Herculis (seen at the bottom of the map) is a wide and easy double—a beautiful object. The components, situated as shown in Plate 3, are of the fourth and eighth magnitude, and coloured respectively greenish-white and grape-red.

The star κ Herculis is not shown in the map, but may be very readily found, lying between the two gammas, γ Herculis and γ Serpentis (see Frontispiece, Map 2), rather nearer the latter. It is a wide double, the components of fifth and seventh magnitude, the larger yellowish-white, the smaller ruddy yellow.*

Ras Algethi, or α Herculis, is also beyond the limits of the map, but may be easily found by means of Map 2, Frontispiece. It is, properly speaking, a multiple star. Considered as a double, the arrangement of the components is that shown in Plate 3. The larger is of magnitude $3\frac{1}{2}$, the smaller of magnitude $5\frac{1}{2}$; the former orange, the latter emerald. The companion stars are small, and require a good telescope to be well seen. Ras Algethi is a variable, changing from magnitude 3 to magnitude $3\frac{1}{2}$ in a period of $66\frac{1}{4}$ days.

The star ρ Herculis is a closer double. The components are 3″·7 apart, and situated as shown in Plate 3. The larger is of magnitude 4, the smaller $5\frac{1}{2}$; the former bluish-white, the latter pale emerald.

There are other objects within the range of our map which are well worthy of study. Such are μ Draconis, a beautiful miniature of Castor; γ¹ and γ² Draconis, a wide double, the distance between the components being nearly 62″ (both grey); and γ¹

---

* They were so described by Admiral Smyth in 1839. Mr. Main, in 1862, describes them as straw-coloured and reddish, while Mr. Webb, in 1865, saw them pale-yellow and *lilac!*

and γ^2 Coronæ, a naked-eye double, the components being 6′ apart, and each double with a good 3-inch telescope.

We turn, however, to another region of the sky. Low down towards the south is seen the small constellation Corvus, recognised by its irregular quadrilateral of stars. Of the two upper stars, the left-hand one is Algorab, a wide double, the components placed as in Plate 3, 23″·5 apart, the larger of magnitude 3, the smaller 8½, the colours pale yellow and purple.

There is a red star in this neighbourhood which is well worth looking for. To the right of Corvus is the constellation Crater, easily recognised as forming a tolerably well-marked small group. The star Alkes, or α Crateris, must first be found. It is far from being the brightest star in the constellation, and may be assumed to have diminished considerably in brilliancy since it was entitled α by Bayer. It will easily be found, however, by means of the observer's star maps. If now the telescope be directed to Alkes, there will be found, following him at a distance of 42·5 s, and about one minute southerly, a small red star, R. Crateris. Like most red stars, this one is a variable. A somewhat smaller blue star may be seen in the same field.

There is another red star which may be found pretty easily at this season. First find the stars η and o Leonis, the former forming with Regulus (now lying towards the south-west, and almost exactly midway between the zenith and the horizon) the handle of the Sickle in Leo, the other farther off from Regulus towards the right, but lower down. Now sweep from o towards η with a low power.* There will be found a sixth-magnitude star about

* Or the observer may sweep from o towards ν, looking for R about two-fifths of the way from o to ν.

one-fourth of the way from o to η. South, following this, will be found a group of four stars, of which one is crimson. This is the star R Leonis. Like R Crateris and R Leporis it is variable.

Next, let the observer turn towards the south again. Above Corvus, in the position shown in the Frontispiece, there are to be seen five stars, forming a sort of wide V with somewhat bowed legs. At the angle is the star γ Virginis, a noted double. In 1756 the components were $6\frac{1}{2}$ seconds apart. They gradually approached till, in 1836, they could not be separated by the largest telescopes. Since then they have been separating, and they are now $4\frac{1}{2}$ seconds apart, situated as shown in Plate 3. They are nearly equal in magnitude (4), and both pale yellow.

The star γ Leonis is a closer and more beautiful double. It will be found above Regulus, and is the brightest star on the blade of the Sickle. The components are separated by about $3\frac{1}{2}$ seconds, the larger of the second, the smaller of the fourth magnitude; the former yellow-orange, the latter greenish-yellow.

Lastly, the star ι Leonis may be tried. It will be a pretty severe test for our observer's telescope, the components being only 2″·4 apart, and the smaller scarcely exceeding the eighth magnitude. The brighter (fourth magnitude) is pale yellow, the other light blue.

CHAPTER IV.

A HALF-HOUR WITH BOOTES, SCORPIO, OPHIUCHUS, ETC.

WE now commence a series of observations suited to the third quarter of the year, and to the following hours:—Ten o'clock on the 22nd of July; nine on the 8th of August; eight on the 23rd of August; seven on the 8th of October; and intermediate hours on days intermediate to these.

We look first for the Great Bear towards the north-west, and thence find the Pole-star. Turning towards the north we see Capella and β Aurigæ low down and slightly towards the left of the exact north point. The Milky Way crosses the horizon towards the north-north-east and passes to the opposite point of the compass, attaining its highest point above the horizon towards east-south-east. This part of the Milky Way is well worth observing, being marked by singular variations of brilliancy. Near Arided (the principal star of Cygnus, and now lying due east—some twenty-five degrees from the zenith) there is seen a straight dark rift, and near this space is another larger cavity, which has been termed the northern Coal-sack. The space between γ, δ, and β Cygni is covered by a large oval mass, exceedingly rich and brilliant. The neighbouring branch, extending from ϵ Cygni, is far less conspicuous here, but near Sagitta becomes brighter than the other, which in this neighbourhood suddenly loses its brilliancy and fading gradually beyond this point becomes invisible near

PLATE IV.

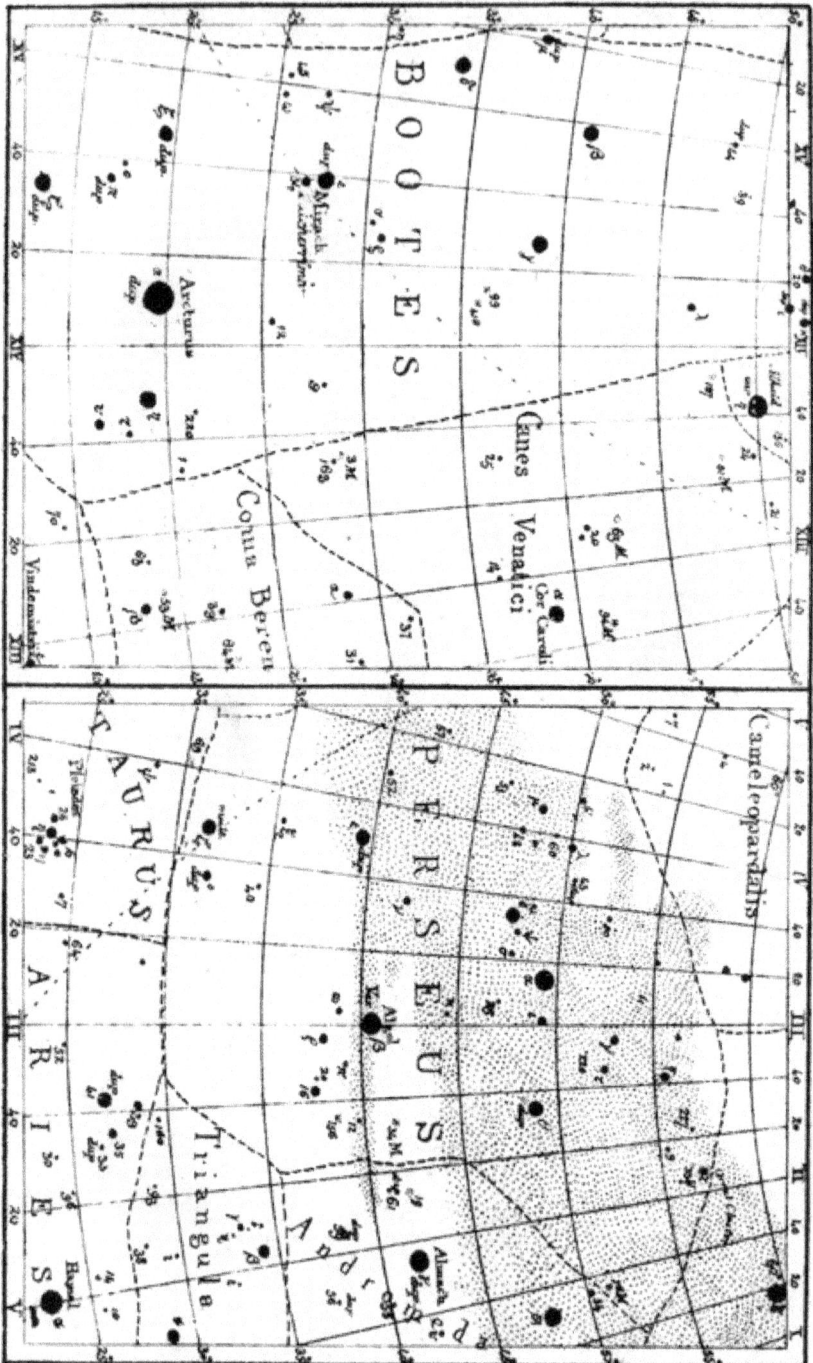

R. A. Proctor, del.¹ et lith.¹

β Ophiuchi. The continuous stream becomes patchy —in parts very brilliant—where it crosses **Aquila** and **Clypeus**. In this neighbourhood the other stream reappears, passing over a region very rich in stars. We see now the greatest extent of the Milky Way, towards this part of its length, ever visible in our latitudes—just as in spring we see its greatest extent towards Monoceros and Argo.

I may note here in passing that Sir John Herschel's delineation of the northern portion of the Milky Way, though a great improvement on the views given in former works, seems to require revision, and especially as respects the very remarkable patches and streaks which characterise the portion extending over Cepheus and Cygnus. It seems to me, also, that the evidence on which it has been urged that the stars composing the Milky Way are (on an average) comparable in magnitude to our own sun, or to stars of the leading magnitudes, is imperfect. I believe, for instance, that the brilliant oval of milky light in Cygnus comes from stars intimately associated with the leading stars in that constellation, and not far removed in space (proportionately) beyond them. Of course, if this be the case, the stars, whose combined light forms the patch of milky light, must be far smaller than the leading brilliants of Cygnus. However, this is not the place to enter on speculations of this sort; I return therefore to the business we have more immediately in hand.

Towards the east is the square of Pegasus low down towards the horizon. Towards the south is Scorpio, distinguished by the red and brilliant Antares, and by a train of conspicuous stars. Towards the west is Bootes, his leading brilliant—the ruddy Arcturus—lying somewhat nearer the horizon than the zenith, and slightly south of west. Bootes as

a constellation is easily found if we remember that
he is delineated as chasing away the Greater Bear.
Thus at present he is seen in a slightly inclined
position, his head (marked by the third-magnitude
star β) lying due west, some thirty degrees from the
zenith. It has always appeared to me, by the way,
that Bootes originally had nobler proportions than
astronomers now assign to him. It is known that
Canes Venatici now occupy the place of an upraised arm of Bootes, and I imagine that Corona
Borealis, though undoubtedly a very ancient constellation, occupies the place of his other arm.
Giving to the constellation the extent thus implied,
it exhibits (better than most constellations) the
character assigned to it. One can readily picture
to oneself the figure of a Herdsman with upraised
arms driving Ursa Major before him. This view is
confirmed, I think, by the fact that the Arabs called
this constellation the Vociferator.

Bootes contains many beautiful objects. Partly
on this account, and partly because this is a constellation with which the observer should be specially familiar, a map of it is given in Plate 4.

Arcturus has a distant pale lilac companion, and
is in other respects a remarkable and interesting
object. It is of a ruddy yellow colour. Schmidt,
indeed, considers that the star has changed colour
of late years, and that whereas it was once very red
it is now a yellow star. This opinion does not
seem well grounded, however. The star *may* have
been more ruddy once than now, though no other
observer has noticed such a peculiarity; but it is
certainly not a pure yellow star at present (at any
rate as seen in our latitude). Owing probably to
the difference of colour between Vega, Capella and
Arcturus, photometricians have not been perfectly
agreed as to the relative brilliancy of these objects.

Some consider Vega the most brilliant star in the
northern heavens, while others assign the supe-
riority to Capella. The majority, however, consider
Arcturus the leading northern brilliant, and in the
whole heavens place three only before him, viz.,
Sirius, Canopus, and α Centauri. Arcturus is remarkable in other respects. His proper motion is
very considerable, so great in fact that since the
time of Ptolemy the southerly motion (alone) of
Arcturus has carried him over a space nearly half
as great again as the moon's apparent diameter.
One might expect that so brilliant a star, apparently
travelling at a rate so great compared with the
average proper motions of the stars, must be comparatively near to us. This, however, has not been
found to be the case. Arcturus is, indeed, one of
the stars whose distance it has been found possible
to estimate roughly. But he is found to be some
three times as far from us as the small star 61
Cygni, and more than seven times as far from us
as α Centauri.

The star δ Bootis is a wide and unequal double,
the smaller component being only of the ninth
magnitude.

Above Alkaid the last star in the tail of the
Greater Bear, there will be noticed three small
stars. These are θ, ι, and κ Bootis, and are usually
placed in star-maps near the upraised hand of the
Herdsman. The two which lie next to Alkaid, ι and
κ, are interesting doubles. The former is a wide
double (see Plate 5), the magnitudes of components
4 and 8, their colours yellow and white. The
larger star of this pair is itself double. The star κ
Bootis is not so wide a double (see Plate 5), the
magnitudes of the components 5 and 8, their colours
white and faint blue—a beautiful object.

The star ξ Bootis is an exceedingly interesting

object. It is double, the colours of the components being orange-yellow and ruddy purple, their magnitudes 3½ and 6½. When this star was first observed by Herschel in 1780 the position of the components was quite different from that presented in Plate 5. They were also much closer, being separated by a distance of less than 3½ seconds. Since that time the smaller component has traversed nearly a full quadrant, its distance from its primary first increasing, till in 1831 the stars were nearly 7½ seconds apart, and thence slowly diminishing, so that at present the stars are less than 5 seconds apart. The period usually assigned to the revolution of this binary system is 117 years, and the period of peri-astral passage is said to be 1779. It appears to me, however, that the period should be about 108 years, the epoch of last peri-astral passage 1777 and of next peri-astral passage, therefore, 1885. The angular motion of the secondary round the primary is now rapidly increasing, and the distance between the components is rapidly diminishing, so that in a few years a powerful telescope will be required to separate the pair.

Not far from ξ is π Bootis, represented in Plate 5 as a somewhat closer double, but in reality—now at any rate—a slightly wider pair, since the distance between the components of ξ has greatly diminished of late. Both the components of π are white, and their magnitudes are 3½ and 6.

We shall next turn to an exceedingly beautiful and delicate object, the double star ε Bootis, known also as Mirac and, on account of its extreme beauty, called Pulcherrima by Admiral Smyth. The components of this beautiful double are of the third and seventh magnitude, the primary orange, the secondary sea-green. The distance separating the components is about 3 seconds, perhaps more; it appears to have been slowly increasing during the past ten or twelve years. Smyth assigns to this system a period of revolution of 980 years, but there can be little doubt that the true period is largely in excess of this estimate. Observers in southern latitudes consider that the colours of the components are yellow and blue, not orange and green as most of our northern observers have estimated them.

A little beyond the lower left-hand corner of the map is the star δ Serpentis, in the position shown in the Frontispiece, Map 3. This is the star which at the hour and season depicted in Map 2 formed the uppermost of a vertical row of stars, which has now assumed an almost horizontal position. The components of δ Serpentis are about $3\frac{1}{2}$ seconds apart, their magnitudes 3 and 5, both white.

The stars θ^1 and θ^2 Serpentis form a wide double, the distance between the components being $21\frac{1}{2}$ seconds. They are nearly equal in magnitude, the primary being $4\frac{1}{2}$, the secondary 5. Both are yellow, the primary being of a paler yellow colour than the smaller star. But the observer may not know where to look for θ Serpentis, since it falls in a part of the constellation quite separated from that part in which δ Serpentis lies. In fact θ lies on the extreme easterly verge of the eastern half of the constellation. It is to be looked for at about the same elevation as the brilliant Altair, and (as to azimuth) about midway between Altair and the south.

The stars α^1 and α^2 Libræ form a wide double, perhaps just separable by the naked eye in very favourable weather. The larger component is of the third, the smaller of the sixth magnitude, the former yellow the latter light grey.

The star β Libræ is a beautiful light-green star to the naked eye; in the telescope a wide double, pale emerald and light blue.

In Scorpio there are several very beautiful objects:—

The star Antares or Cor Scorpionis is one of the most beautiful of the red stars. It has been termed the Sirius of red stars, a term better merited perhaps by Aldebaran, save for this that, in our latitude, Antares is, like Sirius, always seen as a brilliant "scintillator" (because always low down), whereas Aldebaran rises high above the horizon. Antares is a double star, its companion being a minute green star. In southern latitudes the companion of Antares may be seen with a good 4-inch, but in our latitudes a larger opening is wanted. Mr. Dawes once saw the companion of Antares shining alone for seven seconds, the primary being hidden by the moon. He found that the colour of the secondary is not merely the effect of contrast, but that this small star is really a green sun.

The star β Scorpionis is a fine double, the components $13''\cdot1$ apart, their magnitudes 2 and $5\frac{1}{2}$, colours white and lilac. It has been supposed that this pair is only an optical double, but a long time must elapse before a decisive opinion can be pronounced on such a point.

The star σ Scorpionis is a wider but much more difficult double, the smaller component being below the 9th magnitude. The colour of the primary (4) is white, that of the secondary maroon.

The star ξ Scorpionis is a neat double, the components $7''\cdot2$ apart, their magnitudes $4\frac{1}{2}$ and $7\frac{1}{2}$, their colours white and grey. This star is really triple, a fifth-magnitude star lying close to the primary.

In Ophiuchus, a constellation covering a wide space immediately above Scorpio, there are several fine doubles. Among others—

39 Ophiuchi, distance between components $12''\cdot1$,

their magnitudes $5\frac{1}{2}$ and $7\frac{1}{2}$, their colours orange and blue.

The star 70 Ophiuchi, a fourth-magnitude star on the right shoulder of Ophiuchus, is a noted double. The distance between the components about $5\frac{1}{2}''$, their magnitudes $4\frac{1}{2}$ and 7, the colours yellow and red. The pair form a system whose period of revolution is about 95 years.

36 Ophiuchi (variable), distance $5''\cdot2$, magnitudes $4\frac{1}{2}$ and $6\frac{1}{2}$, colours red and yellow.

ρ Opiuchi, distance $4''$, colours yellow and blue, magnitudes 5 and 7.

Between α and β Scorpionis the fine nebula 80 M may be looked for. (Or more closely thus:—below β is the wide double ω^1 and ω^2 Scorpionis; about as far to the right of Antares is the star σ Scorpionis, and immediately above this is the fifth-magnitude star 19.) The nebula we seek lies between 19 and ω, nearer to 19 (about two-fifths of the way towards ω). This nebula is described by Sir W. Herschel as "the richest and most condensed mass of stars which the firmament offers to the contemplation of astronomers."

There are two other objects conveniently situated for observation, which the observer may now turn to. The first is the great cluster in the sword-hand of Perseus (see Plate 4), now lying about $28°$ above the horizon between N.E. and N.N.E. The stars γ and δ Cassiopeiæ (see Map 3 of Frontispiece) point towards this cluster, which is rather farther from δ than δ from γ, and a little south of the produced line from these stars. The cluster is well seen with the naked eye, even in nearly full moonlight. In a telescope of moderate power this cluster is a magnificent object, and no telescope has yet revealed its full glory. The view in Plate 5 gives but the faintest conception

of the glories of χ Persei. Sir W. Herschel tried
in vain to gauge the depths of this cluster with his
most powerful telescope. He spoke of the most
distant parts as sending light to us which must
have started 4000 or 5000 years ago. But it
appears improbable that the cluster has in reality
so enormous a longitudinal extension compared with
its transverse section as this view would imply.
On the contrary, I think we may gather from the
appearance of this cluster, that stars are far less
uniform in size than has been commonly supposed,
and that the mere irresolvability of a cluster is
no proof of excessive distance. It is unlikely that
the faintest component of the cluster is farther
off than the brightest (a seventh-magnitude star)
in the proportion of more than about 20 to 19, while
the ordinary estimate of star magnitudes, applied
by Herschel, gave a proportion of 20 or 30 to 1 at
least. I can no more believe that the components
of this cluster are stars greatly varying in distance,
but accidentally seen in nearly the same direction,
(or that they form an *enormously long system* turned
by accident directly towards the earth), than I
could look on the association of several thousand
persons in the form of a procession as a fortuitous
arrangement.

Next there is the great nebula in Andromeda—
known as "the transcendantly beautiful queen of
the nebulæ." It will not be difficult to find this
object. The stars ϵ and δ Cassiopeiæ (Map 3,
Frontispiece) point to the star β Andromedæ. Almost in a vertical line above this star are two
fourth-magnitude stars μ and γ, and close above ν,
a little to the right, is the object we seek—visible
to the naked eye as a faint misty spot. To tell
the truth, the transcendantly beautiful queen of the
nebulæ is rather a disappointing object in an ordi-

nary telescope. There is seen a long oval or lenticular spot of light, very bright near the centre, especially with low powers. But there is a want of the interest attaching to the strange figure of the Great Orion nebula. The Andromeda nebula has been partially resolved by Lord Rosse's great reflector, and (it is said) more satisfactorily by the great refractor of Harvard College. In the spectroscope, Mr. Huggins informs us, the spectrum is peculiar. Continuous from the blue to the orange, the light there "appears to cease very abruptly;" there is no indication of gaseity.

Lastly, the observer may turn to the pair Mizar and Alcor, the former the middle star in the Great Bear's tail, the latter 15' off. It seems quite clear, by the way, that Alcor has increased in brilliancy of late, since among the Arabians it was considered an evidence of very good eyesight to detect Alcor, whereas this star may now be easily seen even in nearly full moonlight. Mizar is a double star, and a fourth star is seen in the same field of view with the others (see Plate 5). The distance between Mizar and its companion is 14"·4; the magnitude of Mizar 3, of the companion 5; their colours white and pale green, respectively.

CHAPTER V.

A HALF-HOUR WITH ANDROMEDA, CYGNUS, ETC.

OUR last half-hour with the double stars, &c., must be a short one, as we have already nearly filled the space allotted to these objects. The observations now to be made are supposed to take place during the fourth quarter of the year,—at ten o'clock on October 23rd; or at nine on November 7th; or at eight on November 22nd; or at seven on December 6th; or at hours intermediate to these on intermediate days.

We look first, as in former cases, for the Great Bear, now lying low down towards the north. Towards the north-east, a few degrees easterly, are the twin-stars Castor and Pollux, in a vertical position, Castor uppermost. Above these, a little towards the right, we see the brilliant Capella; and between Capella and the zenith is seen the festoon of Perseus. Cassiopeia lies near the zenith, towards the north, and the Milky Way extends from the eastern horizon across the zenith to the western horizon. Low down in the east is Orion, half risen above the horizon. Turning to the south, we see high up above the horizon the square of Pegasus. Low down towards the south-south-west is Fomalhaut, pointed to by β and α Pegasi. Towards the west, about half-way between the zenith and the horizon, is the noble cross in Cygnus; below which, towards the left, we see Altair, and his companions β and γ Aquilæ: while towards the right we see the brilliant Vega.

During this half-hour we shall not confine our-

ι Boötis   κ Boötis   Cor Caroli   ξ Boötis   χ Boötis

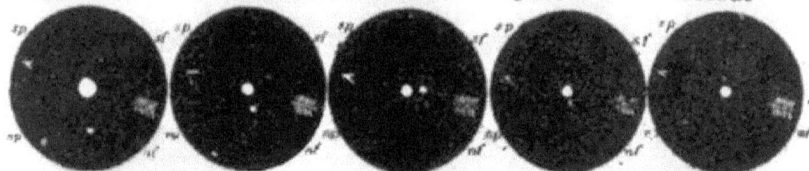

β Scorpii   ι Scorpii   70 Ophuchi   ξ Ophiuchi   δ Serpentis

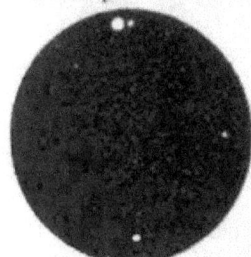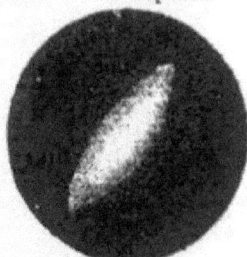

Mizar and Alcor   χ Persei   Gt Neb in Andromeda

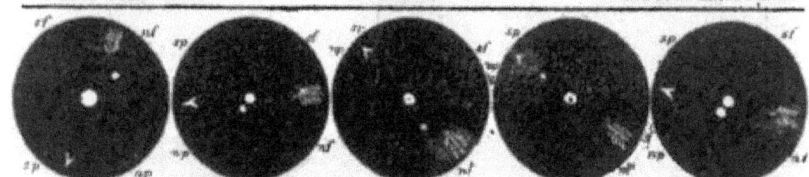

γ Andromeda   α Piscium   ξ Equulei   γ Delphini   ζ Aquarii

Albireo   χ Cygni   61 Cygni   μ Cygni   ψ Cygni

selves to any particular region of the heavens, but sweep the most conveniently situated constellations.

First, however, we should recommend the observer to try and get a good view of the great nebula in Andromeda, which is *not* conveniently situated for observation, but is so high that after a little trouble the observer may expect a more distinct view than in the previous quarter. He will see β Andromedæ towards the south-east, about 18° from the zenith, μ and ν nearly in a line towards the zenith, and the nebula about half-way between β and the zenith.

With a similar object it will be well to take another view of the great cluster in Perseus, about 18° from the zenith towards the east-north-east (*see* the pointers γ and δ Cassiopeiæ in Map 4, Frontispiece), the cluster being between δ Cassiopeiæ and α Persei.

Not very far off is the wonderful variable Algol, now due east, and about 58° above the horizon. The variability of this celebrated object was doubtless discovered in very ancient times, since the name Al-gol, or " the Demon" seems to point to a knowledge of the peculiarity of this "slowly winking eye." To Goodricke, however, is due the rediscovery of Algol's variability. The period of variation is 2 d. 20 h. 48 m. ; during 2 h. 14 m. Algol appears of the second magnitude; the remaining $6\frac{3}{4}$ hours are occupied by the gradual decline of the star to the fourth magnitude, and its equally gradual return to the second. It will be found easy to watch the variations of this singular object, though, of course, many of the minima are attained in the daytime. The following may help the observer :—

On October 8th, 1867, at about half-past eleven in the evening, I noticed that Algol had reached its minimum of brilliancy. Hence the next minimum was attained at about a quarter-past eight on the evening of October 11th; the next at about five on the evening of October 14th, and so on. Now, if this process be carried on, it will be found that the next evening minimum occurred at about 10 h. (*circiter*) on the evening of October 31st, the next at about 11 h. 30 m. on the evening of November 20th. Thus at whatever hour any minimum occurs, another occurs *six weeks and a day later*, at about the same hour. This would be exact enough if the period of variation were *exactly* 2 d. 20 m. 48 s., but the period is nearly a minute greater, and as there are fifteen periods in six weeks and a day, it results that there is a difference of about 13 m. in the time at which the successive recurrences at nearly the same hour take place. Hence we are able to draw up the two following Tables, which will suffice to give all the minima conveniently observable during the next two years. Starting from a minimum at about 11 h. 45 m. on November 20th, 1867, and noticing that the next 43-day period (with the 13 m. added) gives us an observation at midnight on January 2nd, 1868, and that successive periods would make the hour later yet, we take the minimum next after that of January 2nd, viz. that of January 5th, 1868, 8 h. 48 m., and taking 43-day periods (with 13 m. added to each), we get the series—

		h.	m.				h.	m.	
Jan.	5, 1868,	8	45 P.M.	Mar.	10,	1869,	10	25	P.M.
Feb.	17, ——,	8	58 —	Mar.	13,	——,	7	43	—*
Mar.	31, ——,	9	11 —	Apr.	25,	——,	7	56	—
May	13, ——,	9	24 —	June	7,	——,	8	9	—
June	25, ——,	9	37 —	July	20,	——,	8	22	—
Aug.	7, ——,	9	50 —	Sept.	1,	——,	8	35	—
Sept.	19, ——,	10	3 —	Oct.	14,	——,	8	48	—
Nov.	1, ——,	10	16 —	Nov.	26,	——,	9	1	—
Dec.	14, ——,	10	29 —	Jan.	8,	1870,	9	14	—
Jan.	26, 1869,	10	42 —	Feb.	20,	——,	9	27	—

* Here a single period only is taken, to get back to a convenient hour of the evening.

From the minimum at about 10 P.M. on October 31st, 1867, we get in like manner the series—

		h.	m.				h.	m.
Dec.	13, 1867,	10	13 P.M.		Jan.	6, 1869,	8	58 P.M.
Jan.	25, 1868,	10	26 —		Feb.	18, ——,	9	11 —
Mar.	8, ——,	10	39 —		Apr.	2, ——,	9	24 —
Apr.	20, ——,	10	52 —		May	15, ——,	9	37 —
June	2, ——,	11	5 —		June	27, ——,	9	50 —
June	5, ——,	7	53 —*		Aug.	9, ——,	10	3 —
July	18, ——,	8	6 —		Sept.	21, ——,	10	16 —
Aug.	30, ——,	8	19 —		Nov.	3, ——,	10	29 —
Oct.	12, ——,	8	32 —		Dec.	16, ——,	10	42 —
Nov.	24, ——,	8	45 —		Jan.	28, 1870,	10	55 —

From one or other of these tables every observable minimum can be obtained. Thus, suppose the observer wants to look for a minimum during the last fortnight in August, 1868. The first table gives him no information, the latter gives him a minimum at 8 h. 19 m. P.M. on August 30; hence of course there is a minimum at 11 h. 31 m. P.M. on August 27; and there are no other conveniently observable minima during the fortnight in question.

The cause of the remarkable variation in this star's brilliancy has been assigned by some astronomers to the presence of an opaque secondary, which transits Algol at regular intervals; others have adopted the view that Algol is a luminous secondary, revolving around an opaque primary. Of these views the former seems the most natural and satisfactory. It points to a secondary whose mass bears a far greater proportion to that of the primary, than the mass even of Jupiter bears to the sun; the shortness of the period is also remarkable. It may be noticed that observation points to a gradual diminution in the period of Algol's varia-

* Here a single period only is taken, to get back to a convenient hour of the evening.

tion, and the diminution seems to be proceeding more and more rapidly. Hence (assuming the existence of a dark secondary) we must suppose that either it travels in a resisting medium which is gradually destroying its motion, or that there are other dependent orbs whose attractions affect the period of this secondary. In the latter case the decrease in the period will attain a limit and be followed by an increase.

However, interesting as the subject may be, it is a digression from telescopic work, to which we now return.

Within the confines of the second map in Plate 4 is seen the fine star γ Andromedæ. At the hour of our observations it lies high up towards E.S.E. It is seen as a double star with very moderate telescopic power, the distance between the components being upwards of 10''; their magnitudes 3 and 5½, their colours orange and green. Perhaps there is no more interesting double visible with low powers. The smaller star is again double in first-class telescopes, the components being yellow and blue according to some observers, but according to others, both green.

Below γ Andromedæ lie the stars β and γ Trianguiorum, γ a fine naked-eye triple (the companions being δ and η Trianguiorum), a fine object with a very low power. To the right is α Trianguiorum, certainly less brilliant than β. Below α are the three stars α, β, and γ Arietis, the first an unequal and difficult double, the companion being purple, and only just visible (under favourable circumstances) with a good 3-inch telescope; the last an easy double, interesting as being the first ever discovered (by Hook, in 1664), the colours of components white and grey.

Immediately below α Arietis is the star α Ceti,

towards the right of which (a little lower) is Mira, a wonderful variable. This star has a period of $331\frac{1}{3}$ days; during a fortnight it appears as a star of the 2nd magnitude,—on each side of this fortnight there is a period of three months during one of which the star is increasing, while during the other it is diminishing in brightness: during the remaining five months of the period the star is invisible to the naked eye. There are many peculiarities and changes in the variation of this star, into which space will not permit me to enter.

Immediately above Mira is the **star α Piscium at** the knot of the Fishes' connecting band. This is a fine double, the distance between the components being about $3\frac{1}{2}''$, their magnitudes 5 and 6, their colours pale green **and** blue (see Plate 5).

Close to **γ Aquarii** (see Frontispiece, Map 4), above and **to the left of** it, is the interesting double ζ Aquarii; **the distance** between the components is about $3\frac{1}{2}''$, their magnitudes 4 and $4\frac{1}{2}$, both whitish yellow. The period of this binary seems **to be** about 750 years.

Turning next towards the south-west we see the second-magnitude star ε Pegasi, some 40° above the horizon. This star is a wide but not easy double, the secondary being only of the ninth magnitude; its colour is lilac, that of the primary being yellow.

Towards the right of ε Pegasi and lower down **are** seen the three fourth-magnitude stars which mark the constellation Equuleus. Of these the lowest is α, to the right of which lies ε Equulei, a fifth-magnitude star, really triple, but seen as a double star with ordinary telescopes (Plate 5). The distance between the components is nearly $11''$, their colours white and blue, their magnitudes $5\frac{1}{2}$ and $7\frac{1}{2}$. The primary is a very close double, which appears, however, to be opening out rather rapidly.

Immediately below Equuleus are the stars $α^1$ and

a^2 Capricorni, seen as a naked-eye double to the right of and above β. Both a^1 and a^2 are yellow; a^2 is of the 3rd, a^1 of the 4th magnitude; in a good telescope five stars are seen, the other three being blue, ash-coloured, and lilac. The star β Capricorni is also a wide double, the components yellow and blue, with many telescopic companions.

To the right of Equuleus, towards the west-south-west is the constellation Delphinus. The upper left-hand star of the rhombus of stars forming the head of the Delphinus is the star γ Delphini, a rather easy double (see Plate 5), the components being nearly 12″ apart, their magnitudes 4 and 7, their colours golden yellow and flushed grey.

Turn we next to the charming double Albireo, on the beak of Cygnus, about 36° above the horizon towards the west. The components are 34½″ apart, their magnitudes 3 and 6, their colours orange-yellow, and blue. It has been supposed (perhaps on insufficient evidence) that this star is merely an optical double. It must always be remembered that a certain proportion of stars (amongst those separated by so considerable a distance) *must* be optically combined only.

The star χ Cygni is a wide double (variable) star. The components are separated by nearly 26″, their magnitudes 5 and 9, their colours yellow and light blue. χ may be found by noticing that there is a cluster of small stars in the middle of the triangle formed by the stars γ, δ, and β Cygni (see Map 4, Frontispiece), and that χ is the nearest star *of the cluster* to β. The star ϕ Cygni, which is just above and very close to β (Albireo), does not belong to the cluster. χ is about half as far again from ϕ as ϕ from Albireo. But as χ descends to the 11th magnitude at its minimum the observer must not always expect to find it very easily. It has been known to be invisible at the epoch when it should have been most conspicuous. The period of this variable is 406 days.

The star 61 Cygni is an interesting one. So far as observation has yet extended, it would seem to be the nearest to us of all stars visible in the northern hemisphere. It is a fine double, tho components nearly equal ($5\frac{1}{2}$ and 6), both yellow, and nearly $19''$ apart. The period of this binary appears to be about 540 years. To find 61 Cygni note that ϵ and δ Cygni form the diameter of a semicircle divided into two quadrants by α Cygni (Arided). On this semicircle, on either side of α, lie the stars ν and α Cygni, ν towards ϵ. Now a line from α to ν produced passes very near to 61 Cygni at a distance from ν somewhat greater than half the distance of ν from α.

The star μ Cygni lies in a corner of the constellation, rather farther from ζ than ζ from ϵ Cygni. A line from ϵ to ζ produced meets κ Pegasi, a fourth-magnitude star; and μ Cygni, a fifth-magnitude star, lies close above κ Pegasi. The distance between the components is about $5\frac{1}{2}''$, their magnitudes 5 and 6, their colours white and pale blue.

The star ψ Cygni may next be looked for, but for this a good map of Cygnus will be wanted, as ψ is not pointed to by any well-marked stars. A line from α, parallel to the line joining γ and δ, and about one-third longer than that line, would about mark the position of ψ Cygni. The distance between the components of this double is about $3\frac{1}{2}''$, their magnitudes $5\frac{1}{2}$ and 8, their colours white and lilac.

Lastly, the observer may turn to the stars γ_1 and γ_2 Draconis towards the north-west about $40°$ above the horizon (they are included in the second map of Plate 2). They form a wide double, having equal (fifth-magnitude) components, both grey. (See Plate 5.)

CHAPTER VI.

HALF-HOURS WITH THE PLANETS.

In observing the stars, we can select a part of the heavens which may be conveniently observed; and in this way in the course of a year we can observe every part of the heavens visible in our northern hemisphere. But with the planets the case is not quite so simple. They come into view at no fixed season of the year: some of them can never be seen *by night* on the meridian; and they all shift their place among the stars, so that we require some method of determining where to look for them on any particular night, and of recognising them from neighbouring fixed stars.

The regular observer will of course make use of the ʻNautical Almanacʼ; but ʻDietrichsen and Hannayʼs Almanacʼ will serve every purpose of the amateur telescopist. I will briefly describe those parts of the almanac which are useful to the observer.

It will be found that three pages are assigned to each month, each page giving different information. If we call these pages I. II. III., then in order that page I. for each month may fall to the left of the open double page, and also that I. and II. may be open together, the pages are arranged in the following order: I. II. III.; III. I. II.; I. II. III.; and so on.

Now page III. for any month does not concern the amateur observer. It gives information concerning the moon's motions, which is valuable to the sailor, and interesting to the student of astronomy, but not applicable to amateur observation.

PLATE VI.

Mercury.   Venus.

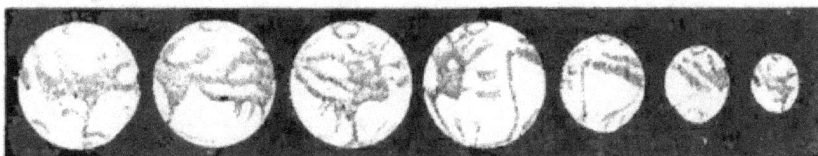

Mars. Summer of the Southern Hemisphere.

Mars. Summer of the Northern Hemisphere.

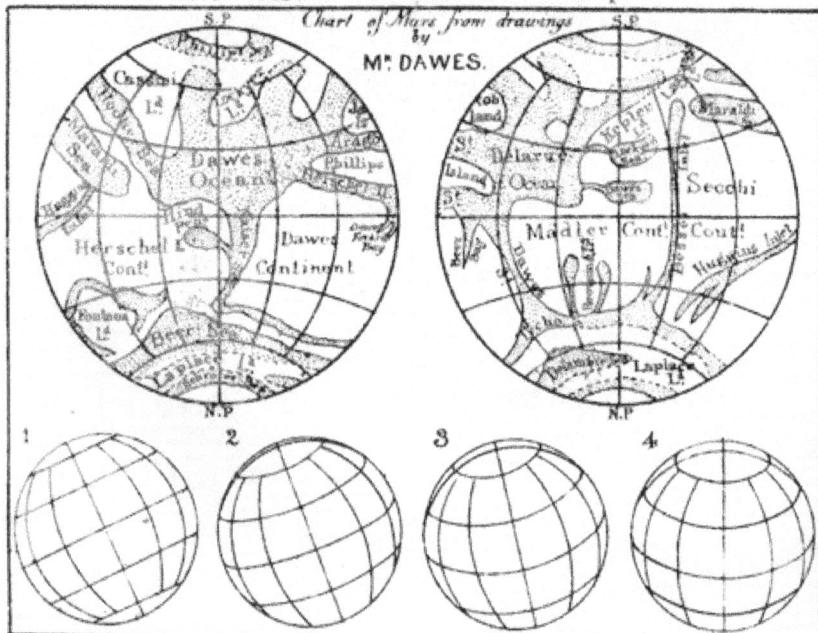

Chart of Mars from drawings by Mr. DAWES.

R. A. Proctor, del. et lith.

We have then only pages I. and II. to consider :—
Across the top of both pages the right ascension
and declination of the planets Venus, Jupiter, Mars,
Saturn, Mercury, and Uranus are given, accompanied
by those of two conspicuous stars. This information
is very valuable to the telescopist. In the first place,
as we shall presently see, it shows him what planets
are well situated for observation, and secondly it
enables him to map down the path of any planet
from day to day among the fixed stars. This is a
very useful exercise, by the way, and also a very
instructive one. The student may either make use
of the regular maps and mark down the planet's path
in pencil, taking a light curve through the points
given by the data in his almanac, or he may lay
down a set of meridians suited to the part of the
heavens traversed by the planet, and then proceed
to mark in the planet's path and the stars, taking
the latter either from his maps or from a convenient list of stars.* My 'Handbook of the Stars'
has been constructed to aid the student in these
processes. It must be noticed that old maps are not
suited for the work, because, through precession, the
stars are all out of place as respects R. A. and Dec.
Even the Society's maps, constructed so as to be
right for 1830, are beginning to be out of date. But
a matter of 20 or 30 years either way is not important.† My Maps, Handbook and Zodiac-chart
have been constructed for the year 1880, so as to be
serviceable for the next fifty years or so.

* I have constructed a zodiac-chart, which will enable the
student to mark in the path of a planet, at any season of the
year, from the recorded places in the almanacs.

† It is convenient to remember that through precession a
star near the ecliptic shifts as respects the R. A. and Dec.
lines, through an arc of one degree — or nearly twice the
moon's diameter—in about 72 years, all other stars through
a less arc.

Next, below the table of the planets, we have a set
of vertical columns. These are, in order, the days
of the month, the calendar—in which are included
some astronomical notices, amongst others the diameter of Saturn on different dates, the hours at which
the sun rises and sets, the sun's right ascension,
declination, diameter, and longitude; then eight
columns which do not concern the observer; after
which come the hours at which the moon rises and
sets, the moon's age; and lastly (so far as the observer is concerned) an important column about
Jupiter's system of satellites.

Next, we have, at the foot of the first page, the
hours at which the planets rise, south, and set; and
at the foot of the second page we have the dates
of conjunctions, oppositions, and of other phenomena,
the diameters of Venus, Jupiter, Mars, and Mercury,
and finally a few words respecting the visibility of
these four planets.

After the thirty-six pages assigned to the months
follow four (pp. 42-46) in which much important
astronomical information is contained; but the points
which most concern our observer are (i.) a small
table showing the appearance of Saturn's rings, and
(ii.) a table giving the hours at which Jupiter's
satellites are occulted or eclipsed, re-appear, &c.

We will now take the planets in the order of their
distance from the sun: we shall see that the information given by the almanac is very important to
the observer.

Mercury is so close to the sun as to be rarely seen
with the naked eye, since he never sets much more
than two hours and a few minutes after the sun, or
rises by more than that interval before the sun. It
must not be supposed that at each successive epoch
of most favourable appearance Mercury sets so long
after the sun or rises so long before him. It would

occupy too much of our space to enter into the circumstances which affect the length of these intervals. The question, in fact, is not a very simple one. All the necessary information is given in the almanac. We merely notice that the planet is most favourably seen as an evening star in spring, and as a morning star in autumn.*

The observer with an equatorial has of course no difficulty in finding Mercury, since he can at once direct his telescope to the proper point of the heavens. But the observer with an alt-azimuth might fail for years together in obtaining a sight of this interesting planet, if he trusted to unaided naked-eye observations in looking for him. Copernicus never saw Mercury, though he often looked for him; and Mr. Hind tells me he has seen the planet but once with the naked eye—though this perhaps is not a very remarkable circumstance, since the systematic worker in an observatory seldom has occasion to observe objects with the unaided eye.

By the following method the observer can easily pick up the planet.

Across two uprights (Fig. 10) nail a straight rod, so that when looked at from some fixed point of view the rod may correspond to the sun's path near the time of observation. The rod should be at right-angles to the line of sight to its centre. Fasten another rod at right angles to the first. From the point at which the rods cross measure off and mark

* Mercury is best seen when in quadrature to the sun, but *not* (as I have seen stated) at those quadratures in which he attains his maximum elongation from the sun. This will appear singular, because the maximum elongation is about 27°, the minimum only about 18°. But it happens that in our northern latitudes Mercury is always *south* of the sun when he attains his maximum elongation, and this fact exercises a more important effect than the mere amount of elongation.

on both rods spaces each subtending a degree as
seen from the point of view. Thus, if the point of
view is $9\frac{1}{2}$ feet off, these spaces must each be 2 inches
long, and they must be proportionately less or greater
as the eye is nearer or farther.

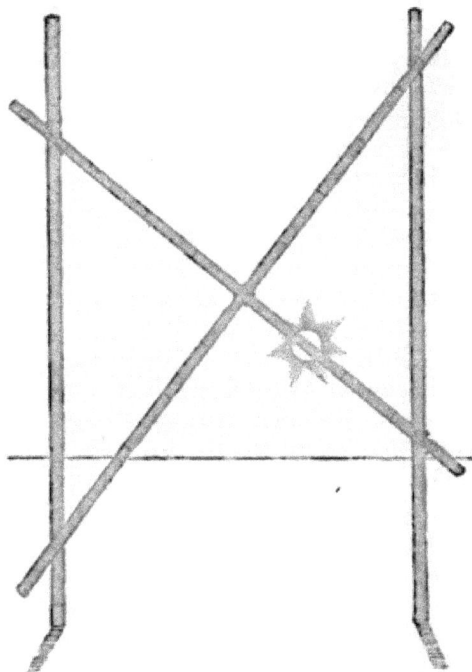

Fig. 10.

Now suppose the observer wishes to view Mercury
on some day, whereon Mercury is an evening star.
Take, for instance, June 9th, 1868. We find from
' Dietrichsen ' that on this day (at noon) Mercury's
R. A. is 6 h. 53 m. 23 s.: and the sun's 5 h. 11 m.
31 s. We need not trouble ourselves about the
odd hours after noon, and thus we have Mercury's

R. A. greater than the sun's by 1 h. 41 m. 52 s. Now
we will suppose that the observer has so fixed his
uprights and the two rods, that the sun, seen from
the fixed point of view, appears to pass the point of
crossing of the rods at half-past seven, then Mercury
will pass the cross-rod at 11 m. 52 s. past nine. But
where ? To learn this we must take out Mercury's
declination, which is 24° 43′ 18″ N., and the sun's,
which is 22° 59′ 10″ N. The difference, 1° 44′ 8″ N.
gives us Mercury's place, which it appears is rather
less than 1¾ degree north of the sun. Thus, about
1 h. 42 m. after the sun has passed the cross-rod,
Mercury will pass it between the first and second
divisions above the point of fastening. The sun will
have set about an hour, and Mercury will be easily
found when the telescope is directed towards the
place indicated.

It will be noticed that this method does not
require the time to be exactly known. All we have
to do is to note the moment at which the sun passes
the point of fastening of the two rods, and to take
our 1 h. 42 m. from that moment.

This method, it may be noticed in passing, may be
applied to give naked-eye observations of Mercury
at proper seasons (given in the almanac). By a little
ingenuity it may be applied as well to morning as to
evening observations, the sun's passage of the cross-
rod being taken on one morning and Mercury's on
the next, so many minutes *before* the hour of the first
observation. In this way several views of Mercury
may be obtained during the year.

Such methods may appear very insignificant to the
systematic observer with the equatorial, but that they
are effective I can assert from my own experience.
Similar methods may be applied to determine from
the position of a known object, that of any neighbouring unknown object even at night. The cross-

rod must be shifted (or else two cross-rods used) when the unknown *precedes* the known object. If two cross-rods are used, account must be taken of the gradual diminution in the length of a degree of right ascension as we leave the equator.

Even simpler methods carefully applied may serve to give a view of Mercury. To show this, I may describe how I obtained my first view of this planet. On June 1st, 1863, I noticed, that at five minutes past seven the sun, as seen from my study window, appeared from behind the gable-end of Mr. St. Aubyn's house at Stoke, Devon. I estimated the effect of Mercury's northerly declination (different of course for a vertical wall, than for the cross-rod in fig. 8, which, in fact, agrees with a declination-circle), and found that he would pass out opposite a particular point of the wall a certain time after the sun. I then turned the telescope towards that point, and focussed for distinct vision of distant objects, so that the outline of the house was seen out of focus. As the calculated time of apparition approached, I moved the telescope up and down so that the field swept the neighbourhood of the estimated point of apparition. I need hardly say that Mercury did not appear exactly at the assigned point, nor did I see him make his first appearance; but I picked him up so soon after emergence that the outline of the house was in the field of view with him. He appeared as a half-disc. I followed him with the telescope until the sun had set, and soon after I was able to see him very distinctly with the naked eye. He shone with a peculiar brilliance on the still bright sky; but although perfectly distinct to the view when his place was indicated, he escaped detection by the undirected eye.*

* It does not seem to me that the difficulty of detecting Mercury is due to the difficulty " of identifying it amongst

Mercury does not present any features of great interest in ordinary telescopes; though he usually appears better defined than Venus, at least as the latter is seen on a dark sky. The phases are pleasingly seen (as shown in Plate 6) with a telescope of moderate power. For their proper observation, however, the planet must be looked for with the telescope in the manner above indicated, as he always shows a nearly semi-circular disc when he is visible to the naked eye.

We come next to Venus, the most splendid of all the planets to the eye. In the telescope Venus disappoints the observer, however. Her intense lustre brings out every defect of the instrument, and especially the chromatic aberration. A dark glass often improves the view, but not always. Besides, an interposed glass has an unpleasant effect on the field of view.

Perhaps the best method of observing Venus is to search for her when she is still high above the horizon, and when therefore the background of the sky is bright enough to take off the planet's glare. The method I have described for the observation of Mercury will prove very useful in the search for Venus when the sun is above the horizon or but just set. Of course, when an object is to be looked for high above the horizon, the two rods which support the cross-rods must not be upright, but square to the line of view to that part of the sky.

But the observer must not expect to see much during his observation of Venus. In fact, he can scarcely do more than note her varying phases (see

the surrounding stars, during the short time that it can be seen " (Hind's 'Introduction to Astronomy'). There are few stars which are comparable with Mercury in brilliancy, when seen under the same light.

G

Plate 6) and the somewhat uneven boundary of the terminator. Our leading observers have done so little with this fascinating but disappointing planet, that amateurs must not be surprised at their own failure.

I suppose the question whether Venus has a satellite, or at any rate whether the object supposed to have been seen by Cassini and other old observers were a satellite, must be considered as decided in the negative. That Cassini should have seen an object which Dawes and Webb have failed to see must be considered utterly improbable.

Leaving the inferior planets, we come to a series of important and interesting objects.

First we have the planet Mars, nearly the last in the scale of planetary magnitude, but far from being the least interesting of the planets. It is in fact quite certain that we obtain a better view of Mars than of any object in the heavens, save the Moon alone. He may present a less distinguished appearance than Jupiter or Saturn, but we see his surface on a larger scale than that of either of those giant orbs, even if we assume that we ever obtain a fair view of their real surface.

Nor need the moderately armed observer despair of obtaining interesting views of Mars. The telescope with which Beer and Mädler made their celebrated series of views was only a 4-inch one, so that with a 3-inch or even a 2-inch aperture the attentive observer may expect interesting views. In fact, more depends on the observer than on the instrument. A patient and attentive scrutiny will reveal features which at the first view wholly escape notice.

In Plate 6 I have given a series of views of Mars much more distinct than an observer may expect to obtain with moderate powers. I add a chart of Mars, a

miniature of one I have prepared from a charming
series of tracings supplied me by Mr. Dawes. The
views taken by this celebrated observer in 1852,
1856, 1860, 1862, and 1864, are far better than any
others I have seen. The views by Beer and Mädler
are good, as are some of Secchi's (though they
appear badly drawn), Nasmyth's and Phillips';
Delarue's two views are also admirable; and Lockyer
has given a better set of views than any of the
others. But there is an amount of detail in Mr.
Dawes' views which renders them superior to any
yet taken. I must confess I failed at a first view to
see the full value of Mr. Dawes' tracings. Faint
marks appeared, which I supposed to be merely
intended to represent shadings scarcely seen. A
more careful study shewed me that every mark is to
be taken as the representative of what Mr. Dawes
actually saw. The consistency of the views is perfectly wonderful, when compared with the vagueness
and inconsistency observable in nearly all other views.
And this consistency is not shown by mere re-
semblance, which might have been an effect rather of
memory (unconsciously exerted) than observation.
The same feature changes so much in figure, as it
appears on different parts of the disc, that it was sometimes only on a careful projection of different views
that I could determine what certain features near the
limb represented. But when this had been done,
and the distortion through the effect of foreshorten-
ing corrected, the feature was found to be as true in
shape as if it had been seen in the centre of the
planet's disc.

In examining Mr. Dawes' drawings it was necessary that the position of Mars' axis should be known.
The data for determining this were taken from
Dr. Oudemann's determinations given in a valuable
paper on Mars issued from Mr. Bishop's observatory.

But instead of calculating Mars' presentation by the formulæ there given, I found it convenient rather to make use of geometrical constructions applied to my 'Charts of the Terrestrial Planets.' Taking Mädler's start-point for Martial longitudes, that is the longitude-line passing near Dawes' forked bay, I found that my results agreed pretty fairly with those in Prof. Phillips' map, so far as the latter went; but there are many details in my charts not found in Prof. Phillips' nor in Mädler's earlier charts.

I have applied to the different features the names of those observers who have studied the physical peculiarities presented by Mars. Mr. Dawes' name naturally occurs more frequently than others. Indeed, if I had followed the rule of giving to each feature the name of its discoverer, Mr. Dawes' name would have occurred much more frequently than it actually does.

On account of the eccentricity of his orbit, Mars is seen much better in some oppositions than in others. When best seen the southern hemisphere is brought more into view than the northern because the summer of his northern hemisphere occurs when he is nearly in aphelion (as is the case with the Earth by the way).

The relative dimensions and presentation of Mars, as seen in opposition in perihelion, and in opposition in aphelion, are shown in the two rows of figures.

In and near quadrature Mars is perceptibly gibbous. He is seen thus about two months before or after opposition. In the former case, he rises late and comes to the meridian six hours or so after midnight. In the latter case, he is well seen in the evening, coming to the meridian at six. His appearance and relative dimensions as he passes from opposition to quadrature are shown in the last three figures of the upper row.

Mars' polar caps may be seen with very moderate powers.

I add four sets of meridians (Plate 6), by filling in which from the charts the observer may obtain any number of views of the planet as it appears at different times.

Passing over the asteroids, which are not very interesting objects to the amateur telescopist, we come to Jupiter, the giant of the solar system, surpassing our Earth more than 1400 times in volume, and overweighing all the planets taken together twice over.

Jupiter is one of the easiest of all objects of telescopic observation. No one can mistake this orb when it shines on a dark sky, and only Venus can be mistaken for it when seen as a morning or evening star. Sometimes both are seen together on the twilight sky, and then Venus is generally the brighter. Seen, however, at her brightest and at her greatest elongation from the sun, her splendour scarcely exceeds that with which Jupiter shines when high above the southern horizon at midnight.

Jupiter's satellites may be seen with very low powers; indeed the outer ones have been seen with the naked eye, and all are visible in a good operaglass. Their dimensions relatively to the disc are shown in Plate 7. Their greatest elongations are compared with the disc in the low-power view.

Jupiter's belts may also be well seen with moderate telescopic power. The outer parts of his disc are perceptibly less bright than the centre.

More difficult of observation are the transits of the satellites and of their shadows. Still the attentive observer can see the shadows with an aperture of two inches, and the satellites themselves with an aperture of three inches.

The minute at which the satellites enter on the disc, or pass off, is given in 'Dietrichsen's Almanac.' The 'Nautical Almanac' also gives the corresponding data for the shadows.

The eclipses of the satellites in Jupiter's shadow, and their occultations by his disc, are also given in 'Dietrichsen's Almanac.'

In the inverting telescope the satellites move from right to left in the nearer parts of their orbit, and therefore transit Jupiter's disc in that direction, and from left to right in the farther parts. Also note that *before* opposition, (i.) the shadows travel in front of the satellites in transiting the disc; (ii.) the satellites are eclipsed in Jupiter's *shadow*; (iii.) they reappear from behind his *disc*. On the other hand, *after* opposition, (i.) the shadows travel *behind* the satellites in transiting the disc; (ii.) the satellites are occulted by the *disc*; (iii.) they reappear from eclipse in Jupiter's *shadow*.

Conjunctions of the satellites are common phenomena, and may be waited for by the observer who sees the chance. An eclipse of one satellite by the shadow of another is not a common phenomenon; in fact, I have never heard of such an eclipse being seen. That a satellite should be quite extinguished by another's shadow is a phenomenon not absolutely impossible, but which cannot happen save at long intervals.

The shadows are not *black spots* as is erroneously stated in nearly all popular works on astronomy. The shadow of the fourth, for instance, is nearly all penumbra, the really black part being quite minute by comparison. The shadow of the third has a considerable penumbra, and even that of the first is not wholly black. These penumbras may not be perceptible, but they affect the appearance of the shadows. For instance, the shadow of the fourth is perceptibly larger but less black than that of the third, though the third is the larger satellite.

In transit the first satellite moves fastest, the fourth slowest, the others in their order. The shadow moves just as fast (appreciably) as the satellite it belongs to. Sometimes the shadow of the satellite may be seen to overtake (apparently) the disc of another. In such a case the shadow does not pass over the disc, but the disc conceals the shadow. This is explained by the fact that the shadow, if visible throughout its length, would be a line reaching slantwise from the satellite it belongs to, and the end of the shadow (that is, the point where it meets the disc) is *not* the point where the shadow crosses the orbit of any inner satellite. Thus the latter may be interposed between the end of the shadow—the only part of the shadow really visible—and the eye; but the end of the shadow *cannot* be interposed between the satellite and the eye. If a satellite *on the disc* were eclipsed by another satellite, the black spot thus formed would be in another place from the black spot on the planet's body. I mention all this because, simple as the question may seem, I have known careful observers to make mistakes on this subject. A shadow is seen crossing the disc and overtaking, apparently, a satellite in transit. It seems therefore, on a first view, that the shadow will hide the satellite, and observers have even said that they have *seen* this happen. But they are deceived. It is obvious that *if one satellite eclipse another, the shadows of both must occupy the same point on Jupiter's body.* Thus it is the overtaking of one *shadow* by another on the disc, and not the overtaking of a *satellite* by a shadow, which determines the occurrence of that as yet unrecorded phenomenon, the eclipse of one satellite by another.[*]

[*] I may notice another error sometimes made. It is said that the shadow of a satellite *appears* elliptical when near

The satellites when far from Jupiter seem to lie in a straight line through his centre. But as a matter of fact they do not in general lie in an exact straight line. If their orbits could be seen as lines of light, they would appear, in general, as very long ellipses. The orbit of the fourth would frequently be seen to be *quite clear* of Jupiter's disc, and the orbit of the third might in some very exceptional instances pass *just* clear of the disc. The satellites move most nearly in a straight line (apparently) when Jupiter comes to opposition in the beginning of February or August, and they appear to depart most from rectilinear motion when opposition occurs in the beginning of May and November. At these epochs the fourth satellite may be seen to pass above and below Jupiter's disc at a distance equal to about one-sixth of the disc's radius.

The shadows do not travel in the same apparent paths as the satellites themselves across the disc, but (in an inverting telescope) *below* from August to January, and *above* from February to July.

We come now to the most charming telescopic object in the heavens—the planet Saturn. Inferior only to Jupiter in mass and volume, this planet surpasses him in the magnificence of his system. Seen in a telescope of adequate power, Saturn is an object of surpassing loveliness. He must be an unimaginative man who can see Saturn for the first time in such a telescope, without a feeling of awe and amazement. If there is any object in the heavens—I except not even the Sun—calculated to impress one with a sense of the wisdom

the edge of the disc. The shadow is *in reality* elliptical when thus situated, but *appears* circular. A moment's consideration will show that this should be so. The part of the disc concealed by a *satellite* near the limb is also elliptical, but of course appears round.

and omnipotence of the Creator it is this. "His fashioning hand" is indeed visible throughout space, but in Saturn's system it is most impressively manifest.

Saturn, to be satisfactorily seen, requires a much more powerful telescope than Jupiter. A good 2-inch telescope will do much, however, in exhibiting his rings and belts. I have never seen him satisfactorily myself with such an aperture, but Mr. Grover has not only seen the above-named features, but even a penumbra to the shadow on the rings with a 2-inch telescope.

Saturn revolving round the sun in a long period —nearly thirty years—presents slowly varying changes of appearance (see Plate 7). At one time the edge of his ring is turned nearly towards the earth; seven or eight years later his rings are as much open as they can ever be; then they gradually close up during a corresponding interval; open out again, exhibiting a different face; and finally close up as first seen. The last epoch of greatest opening occurred in 1856, the next occurs in 1870: the last epoch of disappearance occurred in 1862-63, the next occurs in 1879. The successive views obtained are as in Plate 7 in order from right to left, then back to the right-hand figure (but sloped the other way); inverting the page we have this figure thus sloped, and the following changes are now indicated by the other figures in order back to the first (but sloped the other way and still inverted), thus returning to the right-hand figure as seen without inversion.

The division in the ring can be seen in a good 2-inch aperture in favourable weather. The dark ring requires a good 4-inch and good weather.

Saturn's satellites do not, like Jupiter's, form a system of nearly equal bodies. Titan, the sixth,

is probably larger than any of Jupiter's satellites,
The eighth also (Japetus) is a large body, probably
at least equal to Jupiter's third satellite. But
Rhea, Dione, and Tethys are much less conspicuous,
and the other three cannot be seen without more
powerful telescopes than those we are here dealing
with.

So far as my own experience goes, I consider that
the five larger satellites may be seen distinctly in
good weather with a good 3½-inch aperture. I have
never seen them with such an aperture, but I judge
from the distinctness with which these satellites
may be seen with a 4-inch aperture. Titan is generally to be looked for at a considerable distance
from Saturn—*always* when the ring is widely open.
Japetus is to be looked for yet farther from the
disc. In fact, when Saturn comes to opposition in
perihelion (in winter only this can happen) Japetus
may be as far from Saturn as one-third of the
apparent diameter of the moon. I believe that
under these circumstances, or even under less favourable circumstances, Japetus could be seen with
a good opera-glass. So also might Titan.

Transits, eclipses, and occulations of Saturn's
satellites can only be seen when the ring is turned
nearly edgewise towards the earth. For the orbits
of the seven inner satellites lying nearly in the
plane of the rings would (if visible throughout their
extent) then only appear as straight lines, or as
long ellipses cutting the planet's disc.

The belts on Saturn are not very conspicuous.
A good 3½-inch is required (so far as my experience
extends) to show them satisfactorily.

The rings when turned edgewise either towards
the earth or sun, are not visible in ordinary telescopes, neither can they be seen when the earth and
sun are on opposite sides of the rings. In powerful

telescopes the rings seem never entirely to disappear.

The shadow of the planet on the rings may be well seen with a good 2-inch telescope, which will also show Ball's division in the rings. The shadow of the rings on the planet is a somewhat more difficult feature. The shadow of the planet on the rings is best seen when the rings are well open and the planet is in or near quadrature. It is to be looked for to the left of the ball (in an inverting telescope) at quadrature preceding opposition, and to the right at quadrature following opposition. Saturn is more likely to be studied at the latter than at the former quadrature, as in quadrature preceding opposition he is a morning star. The shadow of the rings on the planet is best seen when the rings are but moderately open, and Saturn is in or near quadrature. When the shadow lies outside the rings it is best seen, as the dark ring takes off from the sharpness of the contrast when the shadow lies within the ring. It would take more space than I can spare here to show how it is to be determined (independently) whether the shadow lies within or without the ring. But the 'Nautical Almanac' gives the means of determining this point. When, in the table for assigning the appearance of the rings, l is less than l' the shadow lies outside the ring, when l is greater than l' the shadow lies within the ring.

Uranus is just visible to the naked eye when he is in opposition, and his place accurately known. But he presents no phenomena of interest. I have seen him under powers which made his disc nearly equal to that of the moon, yet could see nothing but a faint bluish disc.

Neptune also is easily found if his place be accurately noted on a map, and a good finder used. We have only to turn the telescope to a few stars seen in the finder nearly in the place marked in our map, and presently we shall recognise the one we want by the peculiarity of its light. What is the lowest power which will exhibit Neptune as a disc I do not know, but I am certain no observer can mistake him for a fixed star with a 2-inch aperture and a few minutes' patient scrutiny in favourable weather.

PLATE VI.

Jupiter.

Low-power   II.   View

III.   I.   IV.

Saturn.

The Lunar Crater Plato   *Sunrise.*

Plato   *Sunset*

Solar   Spots   From drawings by the
                                                 Rev F Howlett

3

CHAPTER VII.

HALF-HOURS WITH THE SUN AND MOON.

THE Moon perhaps is the easiest of all objects of telescopic observation. A very moderate telescope will show her most striking features, while each increase of power is repaid by a view of new details. Yet in one sense the moon is a disappointing object even to the possessor of a first-class instrument. For the most careful and persistent scrutiny, carried on for a long series of years, too often fails to reward the observer by any new discoveries of interest. Our observer must therefore rather be prepared to enjoy the observation of recognised features than expect to add by his labours to our knowledge of the earth's nearest neighbour.

Although the moon is a pleasing and surprising telescopic object when full, the most interesting views of her features are obtained at other seasons. If we follow the moon as she waxes or wanes, we see the true nature of that rough and bleak mountain scenery, which when the moon is full is partially softened through the want of sharp contrasts of light and shadow. If we watch, even for half an hour only, the changing form of the ragged line separating light from darkness on the moon's disc, we cannot fail to be interested. " The outlying and isolated peak of some great mountain-chain becomes gradually larger, and is finally merged in the general luminous surface; great circular spaces, enclosed with rough and rocky walls many miles in diameter, become apparent; some with flat and perfectly smooth floors, variegated with streaks; others in which the flat floor is dotted with numerous pits or covered with broken fragments of rock. Occasionally a regularly-formed and unusually symmetrical circular formation makes its appearance; the exterior surface of the wall bristling with terraces rising gradually from the plain, the interior one much more steep, and instead of a flat floor, the inner space is concave or cup-shaped, with a solitary peak rising in the centre. Solitary peaks rise from the level plains and cast their long narrow shadows athwart the smooth surface. Vast plains of a dusky tint become visible, not perfectly level, but covered with ripples, pits, and projections. Circular wells, which have no surrounding wall dip below the plain, and are met with even in the interior of the circular mountains and on the tops of their walls. From some of the mountains great streams of a brilliant white radiate in all directions and can be traced for hundreds of miles. We see, again, great fissures, almost perfectly straight and of great length, although very narrow, which appear like the cracks in moist clayey soil when dried by the sun."*

But interesting as these views may be, it was not for such discoveries as these that astronomers examined the surface of the moon. The examination of mere peculiarities of physical condition is, after all, but barren labour, if it lead to no discovery of physical variation. The principal charm of astronomy, as indeed of all observational science, lies in the study of change—of progress, development, and decay, and specially of systematic variations taking place in regularly-recurring cycles. And it is in this relation that the moon has been so disappointing an object of astronomical observation. For two

* From a paper by Mr. Breen, in the 'Popular Science Review,' October, 1864.

centuries and a half her face has been scanned with the closest possible scrutiny; her features have been portrayed in elaborate maps; many an astronomer has given a large portion of his life to the work of examining craters, plains, mountains, and valleys, for the signs of change; but until lately no certain evidence—or rather, no evidence save of the most doubtful character—has been afforded that the moon is other than "a dead and useless waste of extinct volcanoes." Whether the examination of the remarkable spot called Linné — where lately signs were supposed to have been seen of a process of volcanic eruption—will prove an exception to this rule, remains to be seen. The evidence seems to me strongly to favour the supposition of a change of some sort having taken place in this neighbourhood.

The sort of scrutiny required for the discovery of changes, or for the determination of their extent, is far too close and laborious to be attractive to the general observer. Yet the kind of observation which avails best for the purpose is perhaps also the most interesting which he can apply to the lunar details. The peculiarities presented by a spot upon the moon are to be observed from hour to hour (or from day to day, according to the size of the spot) as the sun's light gradually sweeps across it, until the spot is fully lighted; then as the moon wanes and the sun's light gradually passes from the spot, the series of observations is to be renewed. A comparison of them is likely—especially if the observer is a good artist and has executed several faithful delineations of the region under observation, to throw much light upon the real contour of the moon's surface at this point.

In the two lunar views in Plate 7 some of the peculiarities I have described are illustrated. But the patient observer will easily be able to construct for himself a set of interesting views of different regions.

It may be noticed that for observation of the waning moon there is no occasion to wait for those hours in which only the waning moon is visible *during the night.* Of course for the observation of a particular region under a particular illumination, the observer has no choice as to hour. But for generally interesting observations of the waning moon he can wait till morning and observe by daylight. The moon is, of course, very easily found by the unaided eye (in the day time) when not very near to the sun; and the methods described in Chapter V. will enable the observer to find the moon when she is so near to the sun as to present the narrowest possible sickle of light.

One of the most interesting features of the moon, when she is observed with a good telescope, is the variety of colour presented by different parts of her surface. We see regions of the purest white— regions which one would be apt to speak of as *snow-covered,* if one could conceive the possibility that snow should have fallen where (now, at least) there is neither air nor water. Then there are the so-called seas, large grey or neutral-tinted regions, differing from the former not merely in colour and in tone, but in the photographic quality of the light they reflect towards the earth. Some of the seas exhibit a greenish tint, as the Sea of Serenity and the Sea of Humours. Where there is a central mountain within a circular depression, the surrounding plain is generally of a bluish steel-grey colour. There is a region called the Marsh of Sleep, which exhibits a pale red tint, a colour seen also near the Hyrcinian mountains, within a circumvallation called Lichtenburg. The brightest portion of the whole lunar disc is Aristarchus, the peaks of which shine often like stars, when the mountain is within the unillumined portion of the moon. The darkest regions are Grimaldi and Endymion and the great plain called Plato by modern astronomers—but, by Hevelius, the Greater Black Lake.

THE SUN.—Observation of the sun is perhaps on the whole the most interesting work to which the possessor of a moderately good telescope can apply his instrument. Those wonderful varieties in the appearance of the solar surface which have so long perplexed astronomers, not only supply in themselves interesting subjects of observation and examination, but gain an enhanced meaning from the consideration that they speak meaningly to us of the structure of an orb which is the source of light and heat enjoyed by a series of dependent worlds whereof our earth is—in size at least—a comparatively insignificant member. Swayed by the attraction of this giant globe, Jupiter and Saturn, Uranus and Neptune, as well as the four minor planets, and the host of asteroids, sweep continuously in their appointed orbits, in ever new but ever safe and orderly relations amongst each other. If the sun's light and heat were lost, all life and work among the denizens of these orbs would at once cease; if his attractive energy were destroyed, these orbs would cease to form a *system*.

The sun may be observed conveniently in many ways, some more suited to the general observer who has not time or opportunity for systematic observation; others more instructive, though involving more of preparation and arrangement.

The simplest method of observing the sun is to use the telescope in the ordinary manner, protecting the eye by means of dark-green or neutral-tinted glasses. Some of the most interesting views

I have ever obtained of the sun, have resulted from
the use of the ordinary terrestrial or erecting eye-
piece, capped with a dark glass. The magnifying
power of such an eye-piece is, in general, much
lower than that available with astronomical eye-
pieces. But vision is very pleasant and distinct
when the sun is thus observed, and a patient scrutiny
reveals almost every feature which the highest astronomical power applicable could exhibit. Then,
owing to the greater number of intervening lenses,
there is not the same necessity for great darkness or
thickness in the coloured glass, so that the colours
of the solar features are seen much more satisfactorily than when astronomical eye-pieces are
employed.

In using astronomical eye-pieces it is convenient
to have a rotating wheel attached, by which dark-
ening glasses of different power may be brought
into use as the varying illumination may require.

Those who wish to observe carefully and closely
a minute portion of the solar disc, should employ
Dawes' eye-piece: in this a metallic screen placed
in the focus keeps away all light but such as passes
through a minute hole in the diaphragm.

Another convenient method of diminishing the
light is to use a glass prism, light being partially
reflected from one of the exterior surfaces, while the
refracted portion is thrown out at another.

Very beautiful and interesting views may be obtained by using such a pyramidal box as is depicted
in fig. 11.

This box should be made of black cloth or calico
fastened over a light framework of wire or cane.
The base of the pyramid should be covered on the
inside with a sheet of white glazed paper, or with
some other uniform white surface. (Captain Noble,
I believe, makes use of a surface of plaster of Paris,

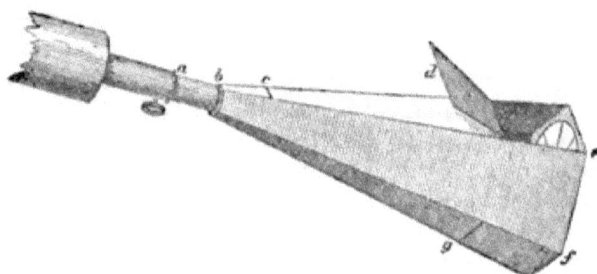

Fig. 11.

smoothed while wet with plate-glass. The door *b c*
enables the observer to " change power" without
removing the box, while larger doors, *d e* and *g f*,
enable him to examine the image; a dark cloth,
such as photographers use, being employed, if necessary, to keep out extraneous light. The image
may also be examined from without, if the bottom
of the pyramid be formed of a sheet of cut-glass or
oiled tissue-paper.

When making use of the method just described,
it is very necessary that the telescope-tube should
be well balanced. A method by which this may be
conveniently accomplished has been already described in Chapter I.

But, undoubtedly, for the possessor of a moderately good telescope there is no way of viewing
the sun's features comparable to that now to be
described, which has been systematically and suc-
cessfully applied for a long series of years by the
Rev. F. Howlett. To use his own words: " Any
one possessing a good achromatic of not more than
three inches' aperture, who has a little dexterity
with his pencil, and a little time at his disposal (all
the better if it be at a somewhat early hour of the
morning)" may by this method "deliberately and
satisfactorily view, measure, and (if skill suffice)

H 2

delineate most of those interesting and grand solar phenomena of which he may have read, or which he may have seen depicted, in various works on physical astronomy." *

The method in question depends on the same property which is involved in the use of the pyramidal box just described, supplemented (where exact and systematic observation is required) by the fact that objects lying on or between the lenses of the eye-piece are to be seen faithfully projected on the white surface on which the sun's image is received. In place, however, of a box carried upon the telescope-tube, a darkened room (or true *camera obscura*) contains the receiving sheet.

A chamber is to be selected, having a window looking towards the south—a little easterly, if possible, so as to admit of morning observation. All windows are to be completely darkened save one, through which the telescope is directed towards the sun. An arrangement is to be adopted for preventing all light from entering by this window except such light as passes down the tube of the telescope. This can readily be managed with a little ingenuity. Mr. Howlett describes an excellent method. The following, perhaps, will sufficiently serve the purposes of the general observer : — A plain frame (portable) is to be constructed to fit into the window : to the four sides of this frame triangular pieces of cloth (impervious to light) are to be attached, their shape being such that when their adjacent edges are sewn together and the flaps stretched out, they form a rectangular pyramid of which the frame is the base. Through the vertex of this pyramid (near

* 'Intellectual Observer' for July, 1867, to which magazine the reader is referred for full details of Mr. Howlett's method of observation, and for illustrations of the appliances he made use of, and of some of his results.

which, of course, the cloth flaps are not sewn together) the telescope tube is to be passed, and an
elastic cord so placed round the ends of the flaps
as to prevent any light from penetrating between
them and the telescope. It will now be possible,
without disturbing the screen (fixed in the window),
to move the telescope so as to follow the sun during
the time of observation. And the same arrangement
will serve for all seasons, if so managed that the
elastic cord is not far from the middle of the telescope-tube; for in this case the range of motion
is small compared to the range of the tube's ex-
tremity.

A large screen of good drawing-paper should next
be prepared. This should be stretched on a light
frame of wood, and placed on an easel, the legs of
which should be furnished with holes and pegs that
the screen may be set at any required height, and
be brought square to the tube's axis. A large
T-square of light wood will be useful to enable the
observer to judge whether the screen is properly
situated in the last respect.

We wish now to direct the tube towards the sun,
and this " without dazzling the eyes as by the ordinary method." This may be done in two ways. We
may either, before commencing work—that is, before
fastening our elastic cord so as to exclude all light
—direct the tube so that its shadow shall be a perfect circle (when of course it is truly directed), then
fasten the cord and afterwards we can easily keep
the sun in the field by slightly shifting the tube as
occasion requires. Or (if the elastic cord has already
been fastened) we may remove the eye-tube and
shift the telescope-tube about — the direction in
which the sun lies being roughly known—until we
see the spot of light received down the telescope's
axis grow brighter and brighter and finally become

a *spot of sun-light*. If a card be held near the focus of the telescope there will be seen in fact an image of the sun. The telescope being now properly directed, the eye-tube may be slipped in again, and the sun may be kept in the field as before.

There will now be seen upon the screen a picture of the sun very brilliant and pleasing, but perhaps a little out of focus. The focusing should therefore next be attended to, the increase of clearness in the image being the test of approach to the true focus. And again, it will be well to try the effect of slight changes of distance between the screen and the telescope's eye-piece. Mr. Howlett considers one yard as a convenient distance for producing an excellent effect with almost any eye-piece that the state of the atmosphere will admit of. Of course, the image becomes more sharply defined if we bring the screen nearer to the telescope, while all the details are enlarged when we move the screen away. The enlargement has no limits save those depending on the amount of light in the image. But, of course, the observer must not expect enlargement to bring with it a view of new details, after a certain magnitude of image has been attained. Still there is something instructive, I think, in occasionally getting a very magnified view of some remarkable spot. I have often looked with enhanced feelings of awe and wonder on the gigantic image of a solar spot thrown by means of the diagonal eye-piece upon the ceiling of the observing-room. Blurred and indistinct through over-magnifying, yet with a new meaning to me, *there* the vast abysm lies pictured ; vague imaginings of the vast and incomprehensible agencies at work in the great centre of our system crowd unbidden into my mind ; and I seem to *feel*—not merely think about—the stupendous grandeur of that life-emitting orb.

To return, however, to observation :—By slightly
shifting the tube, different parts of the solar disc
can be brought successively upon the screen and
scrutinized as readily as if they were drawn upon a
chart. " With a power of—say about 60 or 80 linear
—the most minute solar spot, properly so called,
that is capable of formation" (Mr. Howlett believes
" they are never less than three seconds in length
or breadth) will be more readily detected than by
any other method," see Plate 7 ; "as also will any
faculæ, mottling, or in short, any other phenomena
that may then be existing on the disc." "Drifting
clouds frequently sweep by, to vary the scene, and
occasionally an aërial hail- or snow- storm." Mr.
Howlett has more than once seen a distant flight
of rooks pass slowly across the disc with wonderful
distinctness, when the sun has been at a low altitude, and likewise, much more frequently, the rapid
dash of starlings, which, very much closer at hand,
frequent his church-tower."

An eclipse of the sun, or a transit of an inferior
planet, is also much better seen in this way than by
any other method of observing the solar disc. In
Plate 7 are presented several solar spots as they
have appeared to Mr. Howlett, with an instrument
of moderate power. The grotesque forms of some of
these are remarkable; and the variations the spots
undergo from day to day are particularly interesting
to the thoughtful observer.

A method of measuring the spots may now be
described. It is not likely indeed that the ordinary
observer will care to enter upon any systematic
series of measurements. But even in his case, the
means of forming a general comparison between
the spots he sees at different times cannot fail to
be valuable. Also the knowledge—which a simple
method of measurement supplies—of the actual di-

mensions of a spot in miles (roughly) is calculated
to enhance our estimate of the importance of these
features of the solar disc.   I give Mr. Howlett's
method in his own words :—

"Cause your optician to rule for you on a circular piece of glass a number of fine graduations,
the 200th part of an inch apart, each fifth and tenth
line being of a different length in order to assist the
eye in their enumeration.   Insert this between
the anterior and posterior lenses of a Huygenian
eye-piece of moderate power, say 80 linear.   Direct
your telescope upon the sun, and having so arranged it that the whole disc of the sun may be
projected on the screen, count carefully the number
of graduations that are seen to exactly occupy the
solar diameter. . . . It matters not in which direction you measure your diameter, provided only the
sun has risen some 18° or 20° above the horizon,
and so escaped the distortion occasioned by refraction.*

"Next let us suppose that our observer has been
observing the sun on any day of the year, say, if
you choose, at the time of its mean apparent diameter, namely about the first of April or first of
October, and has ascertained that" (as is the case
with Mr. Howlett's instrument) "sixty-four graduations occupy the diameter of the projected image.
Now the semi-diameter of the sun, at the epochs
above mentioned, according to the tables given for
every day of the year in the 'Nautical Almanac'

---

* As the sun does not attain such an altitude as 18° during
two months in the year, it is well to notice that the true
length of the sun's apparent solar diameter is determinable
even immediately after sun-rise, if the line of graduation is
made to coincide with the *horizontal* diameter of the picture
on the screen—for refraction does not affect the length of
this diameter.

(the same as in Dietrichsen and Hannay's very useful compilation) is 16′ 2″, and consequently his mean total diameter is 32′ 4″ or 1924″. If now we divide 1924″ by 64″ this will, of course, award as nearly as possible 30″ as the value in celestial arc of each graduation, either as seen on the screen, or as applied directly to the sun or any heavenly body large enough to be measured by it."

Since the sun's diameter is about 850,000 miles, each graduation (in the case above specified) corresponds to one-64th part of 850,000 miles—that is, to a length of 13,256 miles on the sun's surface. Any other case can be treated in precisely the same manner.

It will be found easy so to place the screen that the distance between successive graduations (as seen projected upon the screen) may correspond to any desired unit of linear measurement—say an inch. Then if the observer use transparent tracing-paper ruled with faint lines forming squares half-an-inch in size, he can comfortably copy directly from the screen any solar phenomena he may be struck with. A variety of methods of drawing will suggest themselves. Mr. Howlett, in the paper I have quoted from above, describes a very satisfactory method, which those who are anxious to devote themselves seriously to solar observation will do well to study.

It is necessary that the observer should be able to determine approximately where the sun's equator is situated at the time of any observation, in order that he may assign to any spot or set of spots its true position in relation to solar longitude and latitude. Mr. Howlett shows how this may be done by three observations of the sun made at any fixed hour on successive days. Perhaps the following method will serve the purpose of the general observer sufficiently well :—

The hour at which the sun crosses the meridian. must be taken for the special observation now to be described.   This hour can always be learnt from 'Dietrichsen's Almanac'; but noon, civil time, is near enough for practical purposes.   Now it is necessary first to know the position of the ecliptic with reference to the celestial equator.   Of course, at noon a horizontal line across the sun's disc is parallel to the equator, but the position of that diameter of the sun which coincides with the ecliptic is not constant: at the summer and winter solstices this diameter coincides with the other, or is horizontal at noon; at the spring equinox the sun (which travels on the ecliptic) is passing towards the north of the equator, crossing that curve at an angle of $23\frac{1}{2}°$, so that the ecliptic coincides with that diameter of the sun which cuts the horizontal one at an angle of $23\frac{1}{2}°$ and has its *left* end above the horizontal diameter; and at the autumn equinox the sun is descending and the same description applies, only that the diameter (inclined $23\frac{1}{2}°$ to the horizon) which has its *right* end uppermost, now represents the ecliptic.   For intermediate dates, use the following little table :—

Date. (*Circiter.*)	Dec. 22	Jan. 5 / June 6	Jan. 20 / May 21	Feb. 4 / May 5	Feb. 19 / Apr. 20	Mar. 5 / Apr. 5	Mar. 21
Inclination of Ecliptical Diameter of Sun to the Horizon.*	Left 0° 0' Right	Left 6° 24' Right	Left 12° 14' Right	Left 17° 3' Right	Left 20° 36' Right	Left 22° 44' Right	Left 23° 27' Right
Date. (*Circiter.*)	Jan. 21	Dec. 7 / July 7	Nov. 22 / July 23	Nov. 7 / Aug. 6	Oct. 23 / Aug. 23	Oct. 8 / Sept. 7	Sept. 23

* The words "Left" and "Right" indicate which end of the sun's ecliptical diameter is uppermost at the dates in upper or lower row respectively.

Now if our observer describe a circle, and draw
a diameter inclined according to above table, this
diameter would represent the sun's equator if the
axis of the sun were square to the ecliptic-plane.
But this axis is slightly inclined, the effect of which
is, that on or about June 10 the sun is situated as
shown in fig. 14 with respect to the ecliptic *a b* ; on or

Fig. 12. Fig. 13.

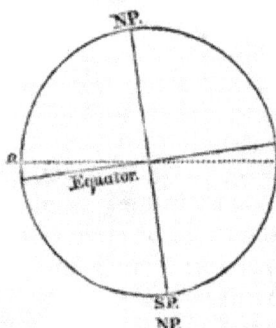
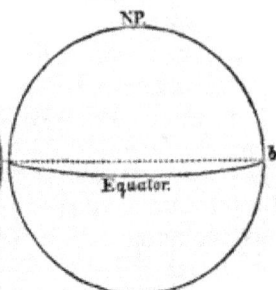

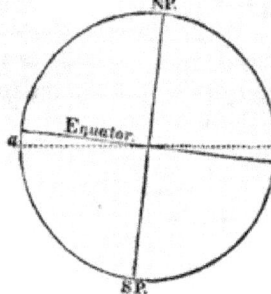
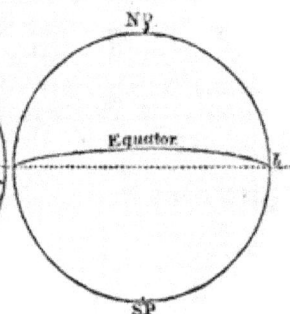

Fig. 14. Fig. 15.

about September 11 he is situated as shown in fig. 13 ;
on or about December 11 as shown in fig. 12 ; and on
or about March 10 as shown in fig. 15. The inclina-
tion of his equator to the ecliptic being so small, the
student can find little difficulty in determining with
sufficient approximation the relation of the sun's

polar axis to the ecliptic on intermediate days, since
the equator is never more *inclined* than in figs. 12
and 14, never more *opened out* than in figs. 13 and
15. Having then drawn a line to represent the
sun's ecliptical diameter inclined to the horizontal
diameter as above described, and having (with this
line to correspond to *a b* in figs. 12—15) drawn in
the sun's equator suitably inclined and opened out,
he has the sun's actual presentation (at noon) as
seen with an erecting eye-piece. Holding his picture upside down, he has the sun's presentation as
seen with an astronomical eye-piece—and, finally,
looking at his picture from behind (without inverting it), he has the presentation seen when the
sun is projected on the screen. Hence, if he make
a copy of this last view of his diagram upon the
centre of his screen, and using a low power, bring
the whole of the sun's image to coincide with the
circle thus drawn (to a suitable scale) on the screen,
he will at once see what is the true position of the
different sun-spots. After a little practice the construction of a suitably sized and marked circle on the
screen will not occupy more than a minute or two.

It must be noticed that the sun's apparent diameter is not always the same. He is nearer to us
in winter than in summer, and, of course, his apparent diameter is greater at the former than at the
latter season. The variation of the apparent diameter corresponds (inversely) to the variation of
distance. As the sun's greatest distance from the
earth is 93,000,000 miles (pretty nearly) and his
least 90,000,000, his greatest, mean, and least apparent diameters are as 93, 91½, and 90 respectively;
that is, as 62, 61, and 60 respectively.

Mr. Howlett considers that with a good 3-inch
telescope, applied in the manner we have described,
all the solar features may be seen, except the sepa-

rate granules disclosed by first-class instruments in the hands of such observers as Dawes, Huggins, or Secchi. Faculæ may, of course, be well seen. They are to be looked for near spots which lie close to the sun's limb.

When the sun's general surface is carefully scrutinised, it is found to present a mottled appearance. This is a somewhat delicate feature. It results, undoubtedly, from the combined effect of the granules separately seen in powerful instruments. Sir John Herschel has stated that he cannot recognise the marbled appearance of the sun with an achromatic. Mr. Webb, however, has seen this appearance with such a telescope, of moderate power, used with direct vision; and certainly I can corroborate Mr. Howlett in the statement that this appearance may be most distinctly seen when the image of the sun is received within a well-darkened room.

My space will not permit me to enter here upon the discussion of any of those interesting speculations which have been broached concerning solar phenomena. We may hope that the great eclipse of August, 1868, which promises to be the most favourable (for effective observation) that has ever taken place, will afford astronomers the opportunity of resolving some important questions. It seems as if we were on the verge of great discoveries,—and certainly, if persevering and well-directed labour would seem in any case to render such discoveries due as man's just reward, we may well say that he deserves shortly to reap a harvest of exact knowledge respecting solar phenomena.

THE END.